T0384212

ORNAMENTAL LIVEBEARERS

This textbook on Ornamental Livebearers is a comprehensive guide and deals with the culture and breeding of livebearers. The present status of ornamental fish farming and new technologies on the breeding and culture of livebearers have also been aptly dealt with. A wide range of aspects such as, anatomy of livebearers, important livebearers and their breeding, feed and feeding management, water quality management, disease management biosecurity and economics of livebearers fish farm have been described in detail. It is hoped that this publication presented in an easy- to- read style with a number of photographs and illustrations would be of great use to all students who have fisheries in their curriculum and also a standard guide for the researchers, entrepreneurs and ornamental fish farmers.

Dr. B. Ahilan, is working as Dean, Dr.MGR Fisheries College and Research Institute, Tamilnadu Dr. Jayalalitha Fisheries University, Ponneri. He has put in 31 years of teaching, research and extension experience in this college. He did his Ph.D dissertation in the breeding and culture goldfish (*Carassius auratus*). He has completed several research projects sponsored by various Central and State government agencies in the areas of Culture and breeding of ornamental fishes, Fish Nutrition and Training of fish farmers.

A. Kamalii is a PG scholar from the Department of Aquaculture, Dr. M.G.R Fisheries College and Research Institute, Ponneri. She is currently working on "Influence of dietary supplementation of black solider fly larvae incorporated with moringa leaf powder on growth performance and colour enhancement of goldfish (*Carassius auratus*) juveniles."

ORNAMENTAL LIVEBEARERS

Dr. B. Ahilan, M.F.Sc., Ph.D
Dean
Dr. M.G.R. Fisheries College & Research Institute
Tamil Nadu Dr. J. Jayalalithaa Fisheries University
Ponneri-601 204, Tiruvallur District, Tamil Nadu

Ms. A. Kamalii, B.F.Sc.
PG Scholar
Department of Aquaculture,
Dr. M.G.R Fisheries College and Research Institute,
Tamil Nadu Dr. J. Jayalalithaa Fisheries University
Ponneri-601 204, Tiruvallur District, Tamil Nadu

NARENDRA PUBLISHING HOUSE
DELHI (INDIA)

First published 2023
by CRC Press
4 Park Square, Milton Park, Abingdon, Oxon, OX14 4RN

and by CRC Press
6000 Broken Sound Parkway NW, Suite 300, Boca Raton, FL 33487-2742

CRC Press is an imprint of Informa UK Limited

British Library Cataloguing-in-Publication Data
A catalogue record for this book is available from the British Library

ISBN: 9781032388946 (hbk)
ISBN: 9781032388953 (pbk)
ISBN: 9781003347323 (ebk)

DOI: 10.4324/9781003347323

Typeset in Arial, Bodoni, Calibri, Cambria, Symbol and Times New Roman
by Amrit Graphics, Delhi 110032

Contents

भा.कृ.अनु.प.– केन्द्रीय मीठाजल जीवपालन अनुसंधान संस्थान
(आईएसओ 9001:2015 प्रमाणित संस्थान)
भारतीय कृषि अनुसंधान परिषद
कौशल्यागंग, भुवनेश्वर–751002, (ओडीसा), भारत
I.C.A.R. - Central Institute of Freshwater Aquaculture
(An ISO 9001:2015 Certified Institute)
Indian Council of Agricultural Research
Kausalyaganga, Bhubaneswar-751002, (Odisha), India

डॉ सरोज कुमार स्वाई
Dr. Saroj K. Swain
निदेशक (कार्यवाहक)
DIRECTOR (Acting)

FOREWORD

Ornamental fishes are referred as living jewels because of their colour, shape, behaviour and origins. They are one among the wonderful creatures of nature, which provide aesthetic beauty to the environment where they are present. There is a lot of interest in commercial farming of ornamental fishes in recent days due to the demand in domestic and international markets. Considerable technical advances in the commercial ornamental fish production have taken place in the last two decades. Suitable technologies for the mass production of egg laying and live bearing fishes have also been developed. Due to these advancements, the world ornamental fish industry has grown considerably and strong value chains were developed across the globe. Many entrepreneurs are getting into this lucrative business enterprise. The livebearers are one of the potent species of culture because of their varied hue and variety of patterns. This Textbook on "Ornamental livebearers" is an excellent piece of work and is the need of the hour too. It will be helpful for the students of Fisheries Science in State Agricultural Universities and in conventional universities where fisheries are taught as a subject. Keeping these in view, this book is brought out in a readable style and at right time. I strongly hope that this publication would be of immense use for the students, farmers and entrepreneurs involved in ornamental aquaculture.

I congratulate the author of this publication for their sincere efforts.

DIRECTOR

Date: 17 June 2021

Phone: 0674 - 2465421, 2465446, 2465502 (O), Fax : 0674 – 2465407
E-mail:director.cifa@icar.gov.in/Saroj.Swain@icar.gov.in, Website:http://www.cifa.nic.in

'Healthy Soils for a Healthy Life'

Preface

A number of publications on Ornamental fishes are available from many countries. But in India, where ornamental fish culture is now being popularized, publications on Ornamental fishes with reference to our agro-climatic conditions are lacking. In order to fill up this long-felt gap, the present work has been taken up.

A wide range of aspects on the culture and breeding of livebearers such as, important livebearers, anatomy of livebearers, Culture and breeding of livebearers feed and feeding management, water quality management, disease management, biosecurity and economics have been dealt with. It is hoped that this contribution presented in an easy to read style with a number of photographs would be of great use to all students who have fisheries in their curriculum and also a standard guide for the researchers and entrepreneurs

We are highly indebted to Dr. Saroj K. Swain, Ph.D., Director, ICAR-Central Institute of Freshwater Aquaculture, Bhubaneswar for his Foreword. We extend our thanks to Mr. A. Kumaran and Ms. D. Arularsi for their secretarial assistance.

Authors

CHAPTER 1 Introduction

Fish is generally regarded as one of the symbols of wealth and good fortune. Feng-shui means wind and water, these two are powerful forces of nature. Aquarium is one of the ways to keep these forces (wind, water) and fishes altogether in beautiful, sparkling manner. The ancient Romans were the first to keep ornamental fishes as pets at home. Ornamental fishes are attractive, vibrant colorful fishes generally regarded as "living jewels". They are not regarded only for their bright colours but also for their peculiar characteristics features such as morphology, mode of taking food, nest building and reproductive strategies etc. which may also add to their wonder.

Aquarium fish keeping began in 1805. The first public aquarium in the world was opened in England in 1852, in the zoological gardens in regrent's park, London and since then keeping and breeding tropical fish has increased in popularity enormously. Ornamental fish keeping has been serving as a viable recreation, especially for hobbyists from time immemorial. It is a second largest hobby next to photography. This sector also plays a valuable role in providing employment opportunities and income to rural, coastal and insular community in the developing world through export generated revenues.

Ornamental fish production globally is a multibillion-dollar industry. The demand for ornamental fish of varied hue and variety is seen increasing rapidly across the world the price of an ornamental fish is considerably higher than the price of a fish destined for human consumption.

Technical advance in this field has taken place in the past 40 years. Recently suitable technologies for mass production of egg-laying and live bearingfishes and breeding operation have been intensified both vertically and horizontally, necessitating a continuous supply of nutritionally, balanced, cost-effective feed. Due to these, the world ornamental fish industry has really grown steadily over the years.

Present Status of Ornamental Fish Culture and Trade

Global Scenario

Ornamental fish keeping is a stress relieving hobby about 7.2 million houses in the USA and 3.24 million in the European union have an aquarium and the number is increasing day by day throughout the world. Meanwhile, ornamental fish farming is also growing to melt out its demand it has been an important economic activity in 125 countries. The global trade value in 2017 was nearly US$ 18-20 billion. The major species involved were neon tetras, angels, goldfish, danios contributing more than14 percent to the global trade. Singapore is the top exporting country in the world. Singapore function by importing and re-exporting the ornamental fishes and accessories and acts as a market hub in Asia. Together all countries of the European union are the largest market for ornamental fish; however, the United States is the single largest importer of ornamental fish in the world (FAO) 1996-2005. In 2017, global ornamental fish import of US was 19.5%. In recent years, ornamental fish production is notably increasing in developing countries like Czech Republic, Malaysia, Thailand and India.

Scenario in India

India is a land of rich 'icthyo fauna biodiversity' and aquarium keeping is the second largest hobby, creating a great demand for the ornamental varieties both in domestic and international market. Nearly 2000 species of ornamental fish varieties are traded annually and the turn of the industry is US$ 7000 million. But the ornamental fish export from India during 2010-2011 was very meager i.e., US$ 1.24 million. Therefore, the ornamental fish industry has a ample scope in India and sky is the limit for it.

The growth of ornamental fish trade inIndia is exotic oriented and is moving towards positive phase. The domestic ornamental fish trade was about Rs. 500 crores. The export trade was Rs 8.40 crores (2017-18) with 11.6% per year growth. The commonly exported fish from India are wild varieties collected from the rivers of northeast and southern states. The north east region contributes 85% to the total market. The culture of fresh water ornamental fishes is mainly limited to the states of West Bengal, Tamil Nadu,Kerala, Maharashtra and recently Karnataka.

West Bengal is the largest ornamental fish producer in India. In Kolkata, ornamental fish farms are located in north and south 24 parganas,Nadia, Hooghly and Howrah districts. Tamilnadu is the second largest ornamental fish producer in the country after West Bengal. Mumbai (Maharashtra) was known for culture ornamental fishes about two-decades back but now it is mainly popular for high value fishes specifically discus only. Kerala ornamental fish culture is recently burgeoning as many villages in the districts of Thiruvanthapuram, Ernakulam, Thrissur, Alappuzha and Kottayam have set up backyard ornamental fish production units.

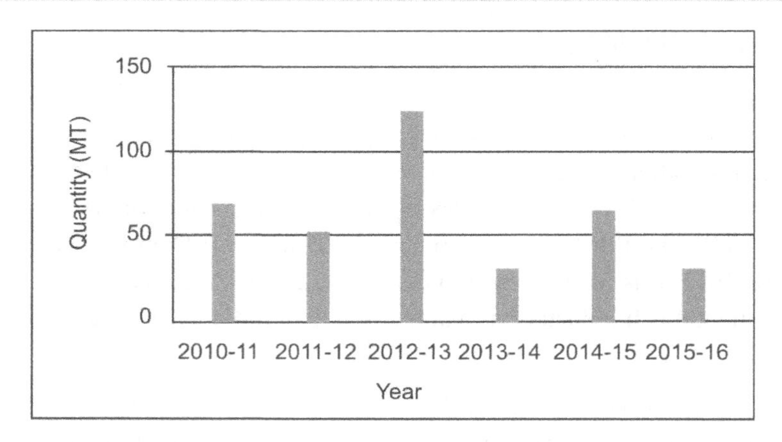

Quantity-wise Ornamental Fish Export from India during 2010-2016 (NFDB, 2017)

PMMSY (Pradhan Mantra Matsya Sampada Yojana)

The Department of fisheries, Ministry of fisheries, animal husbandry and dairying, Government of India is implementing Pradhan Mantra Matsya SampadaYojana – a scheme to bring about blue revolution through sustainable and responsible development of fisheries sector in India at a total investment of Rs 20050 crore for holistic development of fisheries sector including welfare of fishers. PMMSY is implemented in all the States and Union Territories for a period of five years from FY 2020-21 to FY 2024-25. Fishers, fish development of fisheries sector including welfare of fishers.

Under the component for "Development of ornamental and recreation fisheries" they have earmarked Rs 576 crores. The following schemes are implemented for the development of ornamental fisheries:

- Backyard ornamental fish rearing unit (both marine and fresh water with a unit cost of Rs. 3.0lakh.

- Medium scale ornamental fish rearing unit (marine and freshwater fish with a unit cost of Rs. 8.0 lakh.
- Integrated ornamental fish unit (breeding and rearing for freshwater fish) with a unit cost of Rs. 25.0 lakh.
- Integrated ornamental fish unit (breeding and rearing for marine fish) with a unit cost of Rs. 30.0 lakh.

				Governmental Assistance (Rs. Lakhs)	
Sl. No	**Sub-components and activities**	**Unit**	**Unit cost (Rs. Lakhs)**	**General (40%)**	**SC/ST/ Women (60%)**
1	Backyard ornamental fish rearing unit (both marine and freshwater)	Nos	3.2	1.2	1.8
2	Medium Scale Ornamental Fish Rearing Unit (Marine and Freshwater Fish)	Nos	8	3.2	4.8
3	Integrated Ornamental fish unit (Breeding and rearing for fresh water fish)	Nos	25	10	15
4	Integrated Ornamental fish unit (breeding and rearing for marine fish)	Nos	30	12	18
5	Establishment of Fresh water Ornamental Fish Brood Bank	Nos	100	40	60
6	Promotion of Recreational Fisheries	Nos	50	20	30

Scenario in Tamil Nadu

A wide range of availability of species and favourable climate, cheap labour and easy distribution make Tamil Nadu suitable for ornamental fish culture. In Chennai, many farmers grow fish in their backyards and sell the stock to farms, which are engaged in the export business. Ornamental fish culture and trade in TamilNadu especially at kolathur village on the outskirts of Chennai is famous for ornamental culture by small- scale producers. There are about 2000 families earning their livelihood through ornamental fish culture in kolathur and on an average each household in the village earns over Rs. 5,000 to 20,000 per month through ornamental fish farming.

Kolathur approximately generates business worth more than Rs. 60 crore every year.

Shops together sell around 400 species of ornamental fish in wholesale rates, ranging from Rs. 1 to 1.50 lakhs. Besides, there have been several other commercial farms established in different parts of Tamil nadu with emphasis on high valued species.

Pros of Ornamental Fish Keeping

- Ornamental fish keeping gives boundless pleasure to the young and old people.
- It helps in secretions of happy hormones (serotonin, dopamine and endorphins) thus relaxes out mind when we are depressed.
- It keeps the blood pressure at normal level and therefore heart-related disease could be prevented.
- It creates self- employment opportunity.
- It helps to manage the diabetic disease.
- It is good for child development and moreover fish is a good allergy-free animal. It helps the children realize the responsibilities towards the nature and enhances the creativity.

Commercially Important Ornamental Fish Species

About 2000 ornamental fish species have been reported worldwide from various aquatic environments. In India, 217 fish species belonging to 136 genera has been identified in north eastern region, of which about 150 species have been reported to be of ornamental value and in case of more than 50 species, overseas demand has been established. Some of the commercially important egg layers species are gold fish, koi carp, tiger barb, silver shark, red-tailed black shark, zebra danio, rasbora, cardinal tetra, neon tetra, Siamese fighting fish, dwarf gourami, kissing gourami, three-spot gourami, discus, angel fish, Oscar, fire mouthcichlid, Asian arowana, green arowana. Some important live bearer's are guppy, molly, platy and sword tail.

Livebearers

Livebearers are fishes that gives birth to live young ones. The live-bearing fishes clearly evolved from killifish give they birth to live young ones the ultimate is the parental care-giving, sheltering the fertilized eggs within the mother's body during larval development secures a better chance that the baby fish will survive when expelled into the world of predator and prey. Baby livebearers arrive as fully formed miniatures of their parents, ready both to dart to safety and to attack planktonic food. The important ornamental live-bearing fishes are

- Guppy
- Platy
- Molly
- Swordtail

Guppy

Guppy have been around since the early twentieth century. their initial popularity could be because of their tolerance to cooler temperatures and meanwhile can thrive under a wide range of conditions. guppies are of two types:

- Common guppies (seldom exceed an inch and a half in length)
- Fancy guppies (twice as large as common guppies)

Since guppies inhabit moving waters that lack heavy vegetation, we can skip the plantation in the aquarium and simply decorate with rocks and driftwood. However, for breeding purposes plantation should be provided. The plants help the newborn guppies. hide among them and so escapes predation by their cannibalistic parents. The sexes can be easily identified. The males are more colorful and have a gonopodium, by which sperm is introduced into the female. The females are dullin colour.

Platy

Platy the name "platy" is derived from an outdated generic name for these fish, Platypoecilus. The prefix "platy" means "flat" and refers to the much deeper body of these fish compared to the thinner, more elongated members of the livebearer family. All the livebearers are grouped into a single family, the poecillidae. The genetic variability of the platy is evident from the astonishing number of hues that have been developed over the years. They can tolerate wide range of conditions.

Mollies

Mollies (from their old genus name, "mollinesia") are a special subgroup of live-bearing fishes. Earlier, the maintenance of molly was a tedious task as their requirements were not known early. It thrives in the following condition:

- No temperature fluctuations take place
- Diet rich in vegetable matter is given
- Salt is added to the water.

The "green sailfin" mollies, show up now and then in aquarium shops.

Swordtails

The swordtails are closely related to the "platy" fishes. They can be crossbreed. Males are typically slightly smaller than females (6.3 inches) and they reach 5.5 inches maximum. The lifespan of sword tails is up to 5 years. They are omnivorous. They are tolerant to brackish water, so can be kept in low salinities. Sword tails are peaceful species therefore can be placed in a community of other small peaceful fishes.

CHAPTER 2 Types of Ornamental Livebearers

Live bearers are fishes which give birth to live young ones. The different types of ornamental livebearers are:

- Guppy
- Molly
- Sword tail
- Platy

Guppy

Common name	:	Guppy or million fish
Classification	:	Guppy
Phylum	:	Chordata
Super class	:	Pisces
Class	:	Osteichthyes
Sub class	:	Actinopterygii
Order	:	Cyprinodontiformes
Family	:	Poecillidae

Scientific Name : *Poecilia reticulate*
Origin : Central America
Life span : 2 years

Identification

- Guppies are small fishes with a vast variety of caudal fin.
- They have a irregular dorsal fin.
- The mouth is upturned and eyes are large.
- The Guppies have wide variation in their colours.
- Females are larger than the males.
- The maximum size is 8cm.

Sex differentiation

Male	Female
Gonopodium is present	Gonopodium is absent
Bright colouration is identified	Less colouration than the males is identified
The male has a swollen second ray on pelvic fin	The female has no swollen second ray on pelvic fin

Full length: Male: 6cm; female: 8cm

Feeding: They take all type of food

Live food such has blood worm, daphnia, glass worm and tubifex are very good for these fishes

Water quality requirements:

Temperature: 22-24°C

PH: 7.0 to 8.0

Sex ratio: 1:3 (male: female)

Breeding

- They are live bearers since it is cannibalistic in nature breeding traps are provided to save the young ones.
- Courtship begins with behavioural patterns like "Following" the female and "Biting" at her genital region

Community Behavior

Excellent

Fecundity: They give to birth nearly 250 young ones at the intervals between 4-6 weeks.

Feeding of fry: Finely sifted zooplankton, artemia nauplii, may be given, after 14 days occasionally vegetable food can also be given to fry.

The different varieties of Guppy are:

- Black guppy
- Double sword guppy
- Delicate variegated guppy
- Green variegated guppy
- Blue variegated
- Red snakeskin guppy
- Green snakeskin guppy
- Yellow snake skin guppy
- Tuxedo guppy
- Tuxedo cobra guppy
- Red tailed guppy
- Blond red guppy

- Blue tailed guppy
- Pineapple guppy
- Electric blue guppy
- Leopard guppy
- Lyretail guppy

Sword tail

Classification

- Order : Cyprinodontiformes
- Family : Poecillidae
- Scientific name : *Xiphophorus helleri*
- Origin : Central America
- Life span : 5 years

Identification

- The male has a gonopodium
- It has a bright orange and white colour
- The last lower ray of caudal fin is extended like a "sword" like projection, therefore it is known as sword tail fish

Sex differentiation

Male	Female
Gonopodium is present	Gonopodium is absent
If testes develop first, they develop into males by androgen secretion	If ovaries develop first, they develop into females by estrogen secretion

The transformation to female is very rare	They easily turn to a functional male

Full length: Male: 10 cm; Female: 12cm

Feeding: They accept all types of food. Live feeds such as bloodworm, daphnia, glass worm and tubifex are very good for these fish

Water quality requirements:

Temperature: 22°C-25°C

pH: 7.0-7.5

Breeding: They are live bearers.

Fecundity: They give birth to nearly 50-60 young ones at the intervals between 4-6 weeks.

Sex ratio: 1:5 (male: female)

Community behaviour: Excellent

Feeding of fry: Finely sifted zooplankton, artemia nauplii, may be given, after 14 days occasionally vegetable food can also be given to fry.

The different varieties of swordtail are:

- Assorted Swordtail.
- Marigold Wag Swordtail.
- Pineapple Swordtail.
- Red Wag Swordtail.
- Hi Fin Lyretail Swordtail.
- Blood Red Eye Swordtail.

Platy

Common name : Platy

Family : Poecillidae

Scientific name : *Xiphophorus maculatus*

Origin : Central America

Life span : 3-5 years

Identification

- The male can be identified with elongated dorsal fin and a gonopodium.
- Generally, they are orange in colour and black caudal fin.

Sex differentiation

Male	Female
The anal fin of the male fish has evolved into a gonopodium	There is no gonopodium
Caudal fin is more pointed	Caudal fin is fan – shaped (Xiphophorus maculatus)

Full length: Male - 3.5cm; Female - 6.0cm

Feeding: They accept all type of feeds. Live feeds such as bloodworm, daphnia, glass worm and tubifex are very good for these fish.

Water quality requirements:

Temperature: 20-25°C

pH:7.0 – 7.5

Breeding: They are live bearers

Sex ratio: 1:3 (male: female)

Fecundity: Single female gives birth to 30-100 young ones after 20-25 days of mating.

Feeding of fry: They accept fine live as well as artificial foods.

Community behaviour:

- Excellent.
- The other species are
- The southern platy (*Xiphophorus maculatus*)
- The variable platy (Xiphophorus *variatus*)
- The swordtail platy (Xiphophorus *xiphidium*)

Molly

Classification

Order: Cyprinodontiformes

Family: Poeciliidae

Scientific name: *Poecilia sphenops* (black molly or Atlantic molly)

Origin: Central America

Life span: 5 years

Identification

- It possesses jet black colour.
- Females are larger than male.

Sex differentiation

Male	Female
Gonopodium is present	Gonopodium is absent
Dorsal fin is large and tinted orange	Dorsal fin is not larger

Full length : Male : 5cm; Female : 7-11 cm

Feeding: They accept all type of feeds. Live feeds such as bloodworm, daphnia, glass worm and tubifex are very good for these fish.

Water quality requirements

Parameters	Fry	adults
Temperature	22-24°C	24-27°C
pH	7.0-7.5	7.0-7.5

Breeding: Live bearers

Feeding to fry: Fine live as well as artificial feed after fourteen days vegetable foods also can be given

Sex ratio: 2:1 (male: female)

Other Important species

1. *Poecilia formosa (*amazon molly)
2. *Poecilia tatipinna* (sailfin molly)
3. *Poecilia maxicana* (short fined molly)
4. *Poecilia latipinna* (marble molly)

Black guppy

Double sword guppy

Blue tail guppy

Lyre tail guppy

Electric blue guppy

Blue variegated guppy

Leopard guppy

Red tail guppy

Blond red guppy

Red snake skin guppy

Green snakeskin guppy

Yellow snakeskin guppy

Tuxedo guppy

Tuxedo cobra guppy

Pineapple guppy

Amazon molly

Sailfin molly

Short finned molly

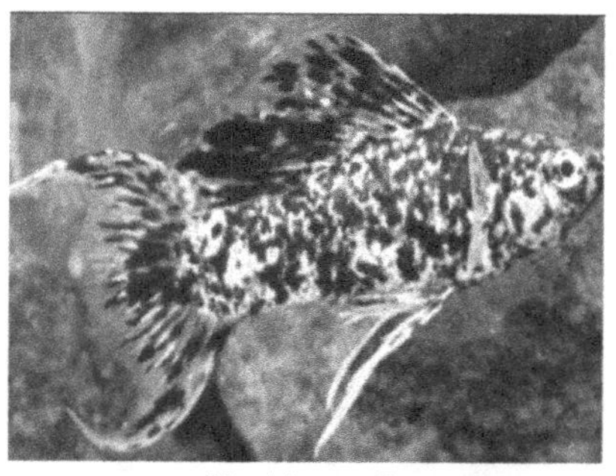

Marble molly

Southern platy

Variable platy

Swordtail platy

Hi fin lyre tail sword tail

Pineapple swordtail

Red wag swordtail

Marigold swordtail

Assorted swordtail

Blood red eye sword tail

CHAPTER 3

Anatomy of Livebearers

Skin and Pigmentation

The skin of livebearers is composed of two layers the thin cellular outer epidermis, and the inner dermis which is composed of connective tissue. The epidermis covers the head and body – scale of the fish, it produces a mucus – it is this which makes the fish feel slimy to the touch –the secretion acting as a protection against diseases and parasites, and it also reduces friction. Beneath the dermis there is a layer of fat (adipose tissue) and muscle (myotomes).

The dermis contains a substance called guanine-a reflective substance that lies just under the scales and also at a lower depth. This light, reflective substance is a white material which collects in clusters of crystals known as iridocytes;

In addition to the iridocytes, the dermis – and to an extend the epidermis – contains colour pigment cells known as chromatophores, these being disposed at different depths within the tissue.

These diverse carotenoids commonly occurring in fishes with their colours are tunaxanthin (yellow), lutein (greenish yellow), canthaxanthin (orange red), astaxanthin (red), echinenone (red), taraxanthin (yellow). The colour of fish skin is primarily dependent on chromatophores (melanophores, xanthophores, erythrophores, iridophores, leucophores, and cyanophores) that contain pigment such as melanin, carotenoids (e.g. astaxanthin, canthaxanthin , lutein , zeaxanthin, pteridines, and purines) Goodwin established that fish do not possess the ability to synthesize carotenoids. The carotenoid pigmentation of fish result from the pigment present in the diet

In fishes there does not exist any universal pathway for metabolism of carotenoids in tissue and its subsequent transformation. It is suggested that organs such as liver or intestine where metabolites of carotenoids exist the metabolism of carotenoids take place.

Colour

Colour is one of the major factors which determines the price of the ornamental fishes in the world market.

The guppy (*Poecilia reticulata*) is thought to be among the most color – polymorphic vertebrates. Male guppies have an outstanding degree of variation in their ornamental patterns, which are shaped by a complex interplay between natural and sexual selection in wild population. Despite our wealth of knowledge about the ecological importance of coloration, the genes and development pathways are underlying.

Guppy pigment pattern formation are unknown, the pigment pattern of the guppy possess of three to four different types of neural crest – derived chromatophores: black melanophores, yellow / orange to reddish xanthophores, blue iridescent iridophores and possibly , white leukophores.

Molly fishes usually has three primitive colors they are:

White, orange and black. Many other colour variations available commercially in the market are:

- The marble with its orange, black and yellow spots

- The red, which is orange but often has pink eyes
- The platinum, which is silver. It is said that it will glow under a black light
- The green, which has a green hue.

The molly is easy to interbreed between different colors that result in offspring with myriad of colour combinations.

In swordtails fishes, the red, orange and black are the most common variants, whereas due to extensive captive breeding we can find so many variant, vibrant colours

Platy colours vary a lot. A few commonly observed colours are yellow, white, blue and green. They interbreed to create more variations in colours and fin.

Gonopodial Diversity

The species of Poecilia (mollies and guppy) and Xiphophorus (swordtails, platy) have very short gonopodia. Therefore, the shorter the gonopodium, the closer a male need to get to a female in order to make contact and effect sperm transfer.

Guppies are internal depositors, with the spermatozeugmata being inserted into the female's genital aperture rather than being laid on its external surface.

In mollies, guppies, swordtail, platies, stabilizing the gonopodium is taken over by the pelvic, rather than pectoral fins. In Xiphophorus (swordtails and platies), males have a "fleshy appendage" along the distal third of the first short, unbranched ray of the pelvic. The second and third rays are somewhat prolonged. In the genus Poecilia (mollies and guppies),the pelvic fins of adult males have a 'variably' developed fleshy swelling distally on ray 1 and a long, fleshy, finger, like extension of the tip of ray 2.

Livebearers exhibit a considerable degree of diversity, both in the overall shape of their bodies. Each permutation is finely tuned to the survival needs of the species in question.\

CHAPTER 4 Aquarium Keeping and Intensive Rearing of Livebearers

Aquarium Keeping of Livebearers

An aquarium is a set up in which ornamental fishes of aesthetic value are kept for recreational activity. The aquaria can be made of myriad of materials like glass, wood, fibreglass acrylic sheet, concrete, etc. The shape of the aquarium can be rectangular, circular, square, oval, hexagonal or octagonal. However, the rectangular tanks are the most preferred one, as they provide an ample amount of swimming space for the fishes. The necessary equipment and accessories required for the aquarium set up are:

- Glass/fibre tank
- Hood
- Stand
- Light source
- Heaters
- Air pumps and accessories
- Filters
- Hand nets
- Monitoring devices
- Feed dispensers
- Cage to hold live-bearers

Aquascaping

The setting up of an aquarium is otherwise called aquascaping. The aquascaping is done for two purposes:

- To make the empty tank attractive for the viewer.
- To mimic the natural environment by keeping aquatic plants, gravels, rocks, etc.

The steps involved in aquascaping are:

- Fitting an under gravel or biological filter to remove the dirty materials from the aquarium.
- The gravels of 3-5mm size are added evenly over the base to foster good water circulation and enable the plants to root. The gravels added should be cleaned properly before added into the aquarium.
- The air pump is installed for the operating purpose of the biological filter for the aeration.

- Large rocks are laid down on the gravel for simulating a natural environmental look.
- The water can be filled with the help of the hosepipe directed over the rock in order to avoid the displacement of the gravels.
- The aquatic plants like sagittaria, vallisneria, java fern, myriophyllum etc. can be introduced in the aquarium which helps in providing hiding space for the fishes and breeding purposes.
- Cover glass or plastic sheet is used to cover the top. These types have cut outs for cables and feeding access. Some of the livebearers like swordtails have jumping behaviour which can be safeguarded by the cover glass.
- Electrical wiring with respect to light and filters are properly connected.
- The aquarium set up is allowed to settle down for the period of 7-14 days to eliminate any excess chlorine or undissolved oxygen in the water and allow plants to establish new roots and anchor themselves to the gravel.
- The hood is installed to which the fluorescent tubes are fitted.
- The newly purchased fishes should not be introduced directly to the tanks. They should be acclimatized properly.
- Immediate feeding of fishes after stocking should be avoided.

Suitability of Live Bearers in Aquarium

Livebearers are great attention seeking fishes because of their vibrant colour and compatibility with other tank mates. They can also be easily bred in the home aquarium.

Habitat and Tank Conditions

The livebearers are hardy species, so they can withstand wide variety of water quality parameters. The livebearers can be kept in aquariums as small as 5 gallons, it depends on the number of fishes being stocked in the aquarium.

- **Guppy:** The general rule is one guppy for every 8 litres of water; therefore, it is 1:8 ratio. For five guppies it requires 40 litres of water.
- **Molly:** The aquarium requires 40 litres of water to fit 4 mollies and in case of sailfin molly and larger molly they require 120 litres of water for 4 number of fishes,
- **Platy:** The aquarium requires 40 litres of water for 5 platy fishes.
- **Swordtail:** One adult swordtail requires at least 60 litres of water as they are active swimmers.

Tankmates

Because of their peaceful nature, they are good community fishes in aquarium tanks. The compatible species consists of peaceful barbs, danios, rasboras, rainbowfish and tetras.

- **Tankmates of guppy:** The best compatible fishes for guppy are platy, swordtail, molly, gourami, Corydoras and ghost shrimp.
- **Tankmates of molly:** The best compatible fishes for molly are guppy, platy, swordtails, cichlids, harlequin rasbora, peaceful barbs, dwarf gourami.
- **Tankmates of platy:** The best compatible fishes for platy are guppy, molly, swordtails, characins, Corydoras, tetras and gourami.
- **Tankmates of swordtail:** The best compatible fishes for swordtail are guppy, platy, molly, Corydoras and angel fish

  It is good to isolate the livebearers from the larger, aggressive fishes.

Intensive Rearing of Livebearers

There are around 100's of millions of ornamental fish sold around the world each year, with more than 80% of these are farm produced species. The usage of conventional systems such as tanks, tubs, etc for ornamental fish farming is being out dated. The advanced burgeoning technologies currently practised to increase the production are:

- LDPE lined ponds
- Raceway ponds
- Recirculatory aquaculture systems

LDPE Lined Ponds

Earthen ponds are provided with lining on sides and bottom with the help of low-density polyethylene (LDPE) or high-density polyethylene (HDPE) sheets. The advantageous factor of lining the pond is that the seepage can be arrested. Further, the pumping cost is considerably reduced in the lined ponds, as we need to fill up the water only to compensate the water loss due to evaporation. The lining materials used are:

LDPE Sheet: LDPE materials are available in different thickness ranging from 150-300 GSM.

HDPE sheets: HDPE sheets are available in different thickness ranging from 500-1000SGM.

The inner lining material for the pond is selected based on the following factors:

- Soil type
- Cost of installation
- Cost of lining material
- Durability
- Suitability

The steps involved in lining the pond with LDPE sheets are:

- The LDPE sheets are cleaned properly.
- Teflon pad will be used beneath the sealing surface for effective sealing.
- The heat in the sealing machine is regulated according to the thickness of the LDPE sheet.
- To avoid the damage due to excess heat a Teflon sheet is also kept over the sheet before sealing.
- After sealing the Teflon sheet can be removed and final check for the effectiveness of the sealing is checked.

Raceway Ponds

A **raceway** which is also known as a **flow-through system**, is an artificial channel used in aquaculture to culture aquatic organisms. A raceway usually consists of rectangular basins or canals constructed of concrete and equipped with an inlet and outlet. A continuous water flow-through is maintained to provide the required level of water quality, which allows animals to be cultured at higher densities within the raceway.

The raceways in green houses are more advantageous compared to the constructional ponds and tanks. The components of the raceway are

- Reservoir pond
- Green house facility
- Pumps, pipes, lines and valves
- Raceway inlets and outlets
- Multiport valves
- Rapid sand filters
- Oxygen injection
- Airlift pumps

Raceway Operation and Management

The raceway is filled with filtered water. The filtered water is allowed to move at a regulated speed, both vertically and horizontally which enables the floating particles feed, faeces, microbes, plankters etc. in suspension and do not settle at the bottom. The principle of the raceway is mechanical agitation and keeping the colloidal particles in suspension which helps in raising the crops successfully.

The water quality and excess nitrogenous waste matter are kept in check by adding probiotics and algal concentrate (as bio remediators). Thus, the intensive rearing of fry can be carried out in this system without affecting the survival and growth rate of the fishes

Advantages of Raceway

- High stocking density can be maintained in this system
- The system can be managed as "zero water exchange system"
- Raceways can adopt "green water technology" to reduce the supplementary feed cost.
- The water quality is managed at optimum levels, which means there is less stress and no disease outbreaks.
- For mass culturing of all kinds of ornamental fishes, the systems are well suited.
- It is an economically viable system.

Recirculatory Aquaculture Systems

RAS is relatively a new technology for ornamental fish production. Increased capital and operating costs are offset by better control over production and higher yields. In a Recirculating Aquaculture System (RAS) the water outflow from the fish tanks is purified and reused. Different treatment units can require different flows and are sometimes operated in a separate loop within the system. Solids are removed by sedimentation or sieving, oxygen is added by aeration or oxygenation, carbon dioxide is removed by degassing and ammonia is mostly converted to nitrate (NO3) by nitrification in aerobic biological filters. Water purification in the treatment units reduces the need for water replacement, so less 'new' water is taken in as the water is more and more re-used. The most basic requirements of a recirculating system are a holding tank for growing and/or holding fish, some form of filtration to maintain water quality, and a pump to recirculate water. An example of a simple recirculating system is a glass aquarium or larger tank with an internal filter.

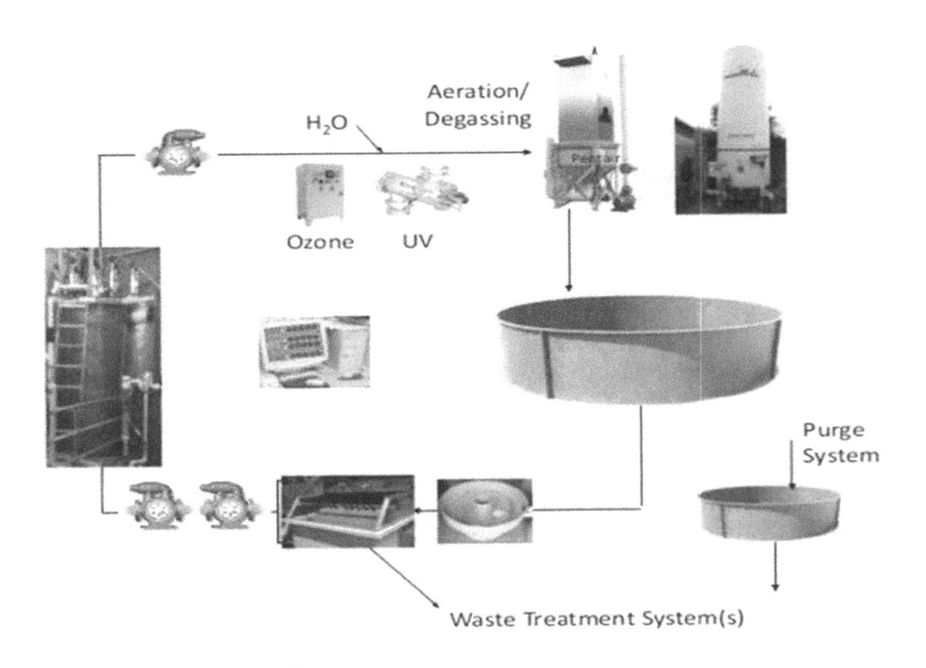

Recirculating Aquaculture System RAS

Species	Commonly used culture systems	Stocking densities	Estimated time to produce market sized fish	Expected Survival (Egg to market size)	Comment
Livebearers	Recirculating systems	1 to 20 fish / l	14-20 weeks	90	Removing females at early age can assist in growing better quality male fish

Disease control is pivotal in RAS, since once the disease has occurred in the RAS it is difficult to eradicate it. The stress and disease are inter-related, control of these factors plays an important role in maintaining the health of the fishes. Stress causes physiological changes that compromise immunity and leaves the fish more susceptible to disease. Maintaining optimal conditions and good hygiene is essential.

CHAPTER 5 Breeding of Livebearers

The fishes bearing young-ones are called livebearers

The livebearers are of two types

- Ovoviviparous
- Viviparous

In ovoviviparous, the egg form and hatch within the female before the birth, where as in viviparous no eggs are formed and the young ones are nourished through an umbilical – like cord (or) from secretion by the female.

Livebearers are prolific breeders and can be bred easily. The important livebearers are guppy, molly, sword tail and platy.

Maturity

Generally, livebearers mature between 4-6 months. However, guppy and platy may mature within two months.

Sex Identification

Characteristic features	Male	Female
Colour	Brightly coloured and attractive	Dull colour
Size	Smaller in size	Larger in size

Fins	Dorsal and caudal fins are longer	Dorsal and caudal fins are comparatively smaller
Belly	Belly region is flat	Belly region bulged
Anal fin	Anal fin is modified into gonopodium	Anal fin is present

Conditioning of Parent Fish

Conditioning is feeding the fish with variety of healthy foods to make them attain suitable condition for spawning.

Before placing the parent fishes together for spawning they should first be conditioned

The parent fishes should be separated while conditioning. Such fishes when reintroduced, they will be ready to spawn.

Breeding of Livebearers

Ideal temperature	-	27°C
pH	-	7.0 - 7.5
Gestation period	-	3 to 4 weeks

In ovoviviparous fishes, the egg is present in the egg duct where they are fertilized upon hatching, the fry is not immediately delivered, but they remain in the safety of the mother's body until they reach a stage of development equivalent to the young of egg layers that have absorbed the yolk sac and become free swimming.

As the male mature, the anal fin is modified into a pointed and straightened rigid tube-like projection, which is called "gonopodium."

The gonopodium is pointed rearwards. However, it is a mobile organ and can be angled in almost any direction courtship may begin with myriad of behavioral patterns, two of which are known as "following" the female and "biting" at her genital region. If the male succeeds during the attention of female, it then commences series of sigmoid display, lending its body into s – shaped curve before the female, so that light reflects the male iridescent colours. During this final performance the male copulate, maneuvering into such a way that he can insert the tip of his gonopodium into the female's genital pore.

Female about to give birth are said to be ripe, this condition can be determined by the appearance of the dark, crescent shaped area in the female body close to the vent known as the gravid spot. The number of young ones in a brood is largely depending upon the size of the female. Irrespective of the number in the brood the fry is approximately all the same size.

The male and the female pail swim in a high circle, with the female flanked by the male, while she arches her hack to expose the genital region, the male transfer bundles of sperm which break up inside the female and migrate to storage pouch near the ovary, where they are retained until required. In guppies, there is development of several young ones at different stages simultaneously a phenomenon known as superfoetation.

Livebearers are notorious cannibals; they will devour their young ones as soon as they are born this can be prevented by using "breeding trap".

Breeding Tank set-up

The preparation for the breeding tank is very easy. the mass breeding requires 100 × 100 × 60cm tank size the depth of 20cm with matured water and the temperature raised to about 25°C – 27°C is ideal for breeding. In case of temperate countries, a water heater is required to maintain the temperature at an optimal level. The tank should be stocked with fine leaved plants. Once the female has given birth to full brood, she can be removed. Handling livebearers when they are near the time of delivery can cause premature birth premature babies have not completely absorbed their yolk sac, which can be seen attached to their bellies.

Live bearers normally kept in a community aquarium tank will breed indiscriminately, and will often crossbreed between similar species. If the aquarist is interested in obtaining a particular colour strain or any other feature for that matter it is imperative that the sexes are housed separately.

Sex ratio (Male: Female)

Guppy – 1:3

Molly – 1:3

Platy – 1:3

Sword tail – 1:5

Number of Young Ones

The number of young-ones produced varies from fish to fish, feed given and tank environment on an average 20-40 young ones can be expected.

Guppies – maximum of 100 young ones

Molly – maximum of 100 young ones

Platy – maximum of 50 young ones

Swordtail – maximum of 200 young-ones

After 2-3 days the female again becomes pregnant even when there is no contact with male. the sperm transferred during first matting is stored in the female body and when eggs are formed, the sperm will join with eggs and form young-ones. Thus, by single mating young-ones can be released 5-6 times. in a year, it is possible to get young ones 10 times from one female.

Feeding of Fry

The fry can swim and eat from the moment they are born. The fry accepts finely sifted zooplankton brine shrimp nauplii, may be given, after 14 days occasional vegetable food can also be given to fry.

Breeding of Guppy

Guppy is one of the most favourite ornamental species preferred by the hobbyist because of their vibrant colours and they can be easily bred without any great deal of expense and knowledge about them. Providing proper water conditions, natural-looking environment and proper filtration will eventually lead to good spawning behaviour of the fishes. As, guppies are ovoviviparous they fertilize internally and give birth to fully formed young ones. The male guppies have sex organ, the gonopodium present in their anal fin. The gonopodium tip consists of several hooks that helps the male to hold the female during the copulation process. The female guppies have a remarkable ability to give birth to young ones consecutively without the male insemination occurring each time, this is because of their ability to store the sperm in the special folds of female genital area until it is later needed for fertilization. Both linebreeding and inbreeding methods can be used to produce pure strains of guppies. The mature male and female with desired characteristic features are taken and kept in the breeding tank containing filters and floating plants. The floating plants provides the hiding space for the newly hatched young ones. Immediately, after the release of young ones the parental fishes should be removed because of their cannibalistic behaviour.

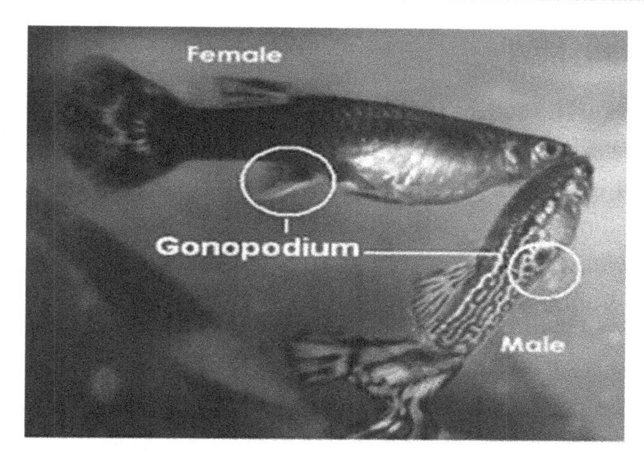

Breeding of Platy

Platy fishes are small peaceful fishes. They are very active; it is indeed necessary to cover the top of the aquarium as they have jumping behaviour. They are prolific breeders. They become sexually mature at about 4 months. The males possess a gonopodium, while the females are identified by their dull colour. The optimal water conditions, floating plants and the presence of male and female platies are perfect for the onset of the breeding process. The pregnant females are identified by the abdominal enlargement. Occasionally, a gravid spot may appear in the females.

Breeding of Molly

Molly fishes will readily breed even in home aquaria if optimal water conditions are available. They male and female mollies in the breeding tank should be provided with adequate space and densely planted live plants for hiding purposes. The live plants apart from providing space for hiding they are excellent source of algae to feed on for the adult fish as well as the young ones.

Breeding of Swordtail

The sword tail fishes are known for their elongated lower lobe on the tail-this extension looks like a sword. The male fishes only posses this sword

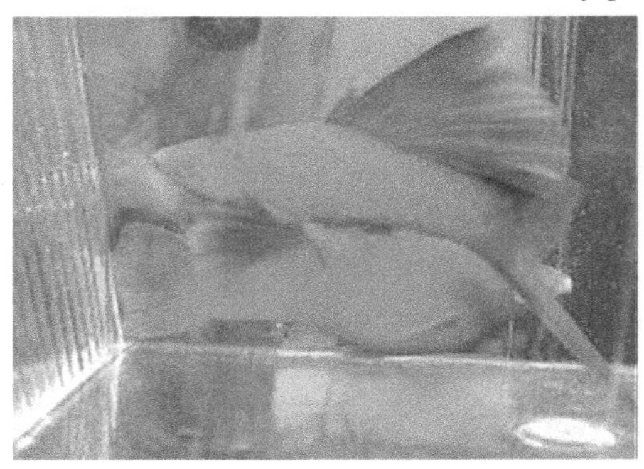

and therefore it is easy for sex differentiation. The swordtails usually breed when there is optimal water conditions and healthy feed is provided. Once the males are ready to breed, they swim alongside the females, nipping their fins. This condition, can induce stress to the females, so providing more females than the males in tanks can reduce this occurrence. The pregnant female, will posses a swollen belly with a gravid spot on the anal fin. The fry should be separated from the parents because of their cannibalistic behaviour.

Species	Sexual dimorphism	Size	Optimum water condition for breeding	Gestation period	Young ones per female	Starter diet
Guppy	Male is smaller with more flowing fins and pointed anal fin or gonopodium	Male:2.5-3.5 cm Female:5-6cm	Temp: 20-28°C, Water hardness 50-100 $CaCO_3$/litre	21-35	20-100	Finely powdered dried feed and rotifers
Platy	Male is smaller and slimmer with gonopodium; Colouration Red, gold, blue, black, brown	Male:3-4cm Female:4-5cm	Temp:23-28°C, Water hardness 50-100 CaCO3/litre	28-42	10-50	Finely powdered dried feed and rotifers

[Table Contd.

Contd. Table]

Species	Sexual dimorphism	Size	Optimum water condition for breeding	Gestation period	Young ones per female	Starter diet
Sword tail	Male is smaller and slimmer with gonopodium and sword like projection on caudal fin	Male: 6- 7cm Female: 7- 9cm	Temp:23- 28°C, Water hardness 50-100$CaCO_3$/ litre	28-42	20-100	Finely powdered dried feed and rotifers
Molly	Male is smaller and slimmer with gonopodium dorsal fin is flowing and bigger	Male:7-8cm Female: 9cm	Temp:23-28°C, Aquariumsalt @0.5-1.0g /litre	40-70	10-100	Finely powdered dried feed and rotifers

CHAPTER 6 Food and Feeding

Feed is a prerequisite factor in determining the growth rate, health and survival of the livebearers in the aquarium. In their natural habitats fishes consume a vast variety of items which includes aquatic organisms, vegetation etc. Some are opportunistic feeders, eating anything that is available in their surrounding and they possess a generalized digestive system to survive in that condition. In case of obligatory feeders, they have evolved intestinal tracts suited to digest only the appropriate diet.

Livebearers are omnivorous and they are in need of both vegetable matter and meat-based foods and they require protein, fat, carbohydrates, minerals and vitamins in their food at appropriate proportions to stay healthy and have an efficient growth rate.

The different types of feed available are

- Live feed
- Artificial feed

Live Feed

Live feed which are commonly known as "living capsules of nutrition" contain all the nutrients such as essential proteins, lipids, carbohydrates, vitamins, minerals, amino acids and fatty acids. Live feed like tubifex, daphnia, moina, earthworm, bloodworm and artemia are regarded by most aquarists as the best form of food, as they are the natural foods that would be taken by the fishes in the wild.

Brine shrimp (Artemia)

Of all the live food used for larvae culture of finfish and shellfish in aqua-hatcheries, artemia is the most widely used organism. The main advantage of using artemia is that, one can produce live food "on demand" through hydration of dormant cysts into nutritious nauplii within 24 hours.

Artemia cysts measures about 200 microns in diameter and are brown in colour. It remains covered with hard shell chorion. Freshly hatched artemia nauplii are about 0.4 mm in length. It is orange in colour due to presence of yolk, containing all essential fatty acids, amino-acids and significant concentration of vitamin, hormones and carotenoids in addition to protein content of nearly 50 precent.

Artemia habitats in salt pans in different periods of the year, which is dependent on salinity conditions. It occurs only in specific areas for certain period. It occurs mainly in high saline water.

It prefers to feed on *Dunaliella* and *Scenedesmus* but cannot digest *Chlorella* due to thick cell wall. *Artemia salina* attains sexual maturity in two weeks' time after hatching and reproduces continuously throughout its life span for 30 to 40 days. Artemia occurs in two stains, one is bisexual (presence of male and female) and the other parthenogenetic (only female).

Hatching of Artemia Cysts

Hydration of Cyst

The required number of dry cysts are put in a container having sea water (or) freshwater, provided with aeration. In an hour, the cysts get hydrated and turns spherical. Examine the cyst under stereoscope before proceeding. The hydrated cyst is filtered in 100μm mesh bolting.

Decapsulation of Cyst

Decapsulation of hydrated cysts of artemia is achieved by heating them with sodium hypochlorite (NaOCl) (or) bleaching powder.

A) **Using sodium hypochlorite:** The hydrated cysts are kept in 5% of sodium hypochlorite @ 15ml for everyone gram cyst, 100g cyst can be used. In order to prevent damage of embryo from raising temperature, keep the cysts in trough containing cool water. In order to remove residual chlorine 0.1% sodium thiosulphate 10.1N HCL is used.

B) **Using bleaching powder:** Bleaching powder is dissolved to get 200 – 250 ppm of chlorine. The hydrated cyst are transferred to the bucket containing bleaching powder and left for 15- 20 mins

Hatching of the Cyst

The optimum water quality conditions required are:

Temperature – 27 -30°C

pH – 7.5 -8.5

Salinity – 25 – 30ppt

Light – 1000lux

Do - saturation point

The jar is filled with 25 -30ppt filtered seawater. Vigorous aeration is given with fluorescent light on top of the jar. Cysts will hatch within 12-14 hours depending on temperature, salinity and stain of artemia.

Earthworms

Earthworms are excellent live feed for the live bearers. They can be easily collected from a garden (or) patch of waste ground almost in most of the seasons in a year. It is easily palatable and acts as a mild laxative. The collected earthworm should be swilled to remove any soil (or) slimy mud which adheres to the earthworm. Depuration should be carried out to wash out solid and waste particles within them and make them more palatable for the fish. The worms can be given as whole (or) chopped, depending on the size of the fishes.

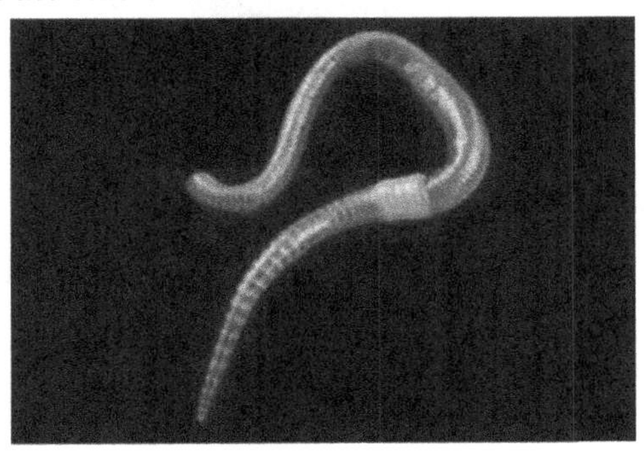

Rotifers

Rotifers are microscopic organism, commonly known as "wheel animalcules" among the rotifer species, *Brachionus* sp are best known for its high nutritive value, small size, fast multiplication, easy adaptability and worldwide distribution.

B. plicatilis is fastidious filter feeder. It prefers food items which are simple in shape and well suspended in water. It obtains food through ciliated corona

The most acceptable food for *B. plicatilis* is *Chlorella* sp. and *Tetraselmis* spp.

B. plicatilis undergo two types of reproduction depending upon the culture condition. Under favourable conditions, the most predominant mode of reproduction is parthenogenesis (amitotic reproduction). However, in unfavorable condition, it restores to sexual reproduction (mystic reproduction.)

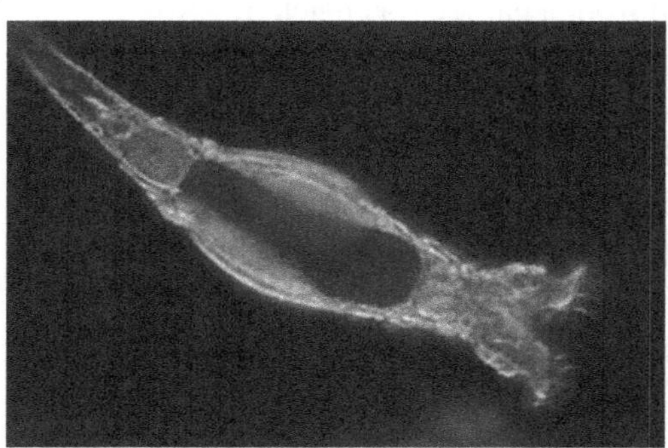

Culture: Stock Culture

In order to start stock, culture collect *B.plicatilis* from the stagnant salt water bodies (salinity between 10 to 40 ppt) with the help of a scoop net having 50-100 micron mesh. Dilute the sample by adding freshwater having same salinity.

Serially dilute the test tube culture daily to test tubes of 20ml capacity containing 10ml water.. gradually increase volume to 50 to 100 ml capacity beaker.

Cladocerans (Daphnia and Moina)

Cladocerans are commonly known as water fleas, their size varies from 0.5 to 1.5mm. the commonly cultured Cladocerans are daphnia and moina, they both have similar culture techniques

Daphnia have a transparent chitinous carapace and vary in size according to age. They breathe through gills and deprive fish of essential oxygen, so do not feed too many at a time. Daphnia feeds on algae, bacteria, fungal, protozoan and organic debris. Daphnia is differentiated from moina by the presence of prominent caudal spire. Daphnia and moina reproduce parthenogenetically.

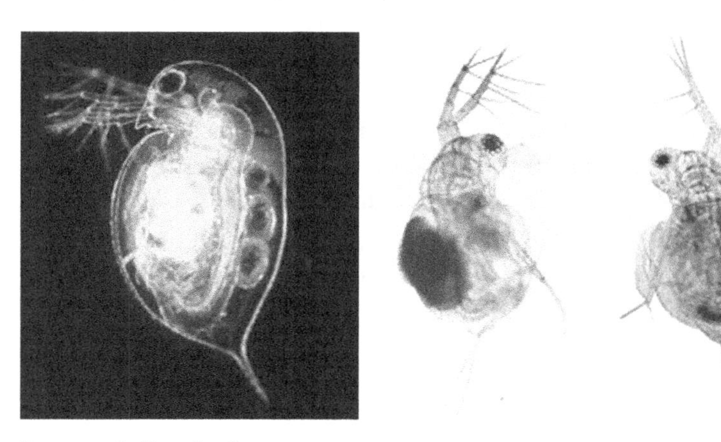

Culture of Daphnia

a) **Stock and pure culture:** Daphnia can be cultured in mass scale level. To prepare stock and pure culture daphnia should be collected and the sample has to be diluted and taken in a glass beaker.

Individual daphnia is picked with the help of a dropper and placed one in each tube containing 10ml of filtered freshwater. Daphnia is fed with yeast or groundnut oil cake at 200ppm daily. This tube culture is inoculated in mass culture tanks

b) **Mass culture**

Depending on the requirements, daphnia is cultured in 500 to 20,000 litre capacity cement or plastic tanks. The culture tanks is thoroughly cleaned and filled with filtered freshwater. Before starting a mass culture, medium is prepared as follows:

Medium preparation: Slurry is prepared by adding 10 kg of chicken droppings, 5kg of groundnut oil cake and 2.5 kg of single super phosphate in 250 litre fresh water. Continuous aeration is given for 3 days for the escape of obnoxious gases, fermentation and release of nutrients. After 3days the slurry is used as fertilization solution in the mass culture tanks, the medium is added at 3 to 4ml per litre of water regularly for 3 to 4 days on 4th day, daphnia is inoculated at 50 individuals per litre. In about 7days, daphnia multiplies and density reaches from 10,000 to25,000 individuals per litre. The daphnia is harvested using 100-200 micron mesh scoop net in the early morning or late evening when they are on water surface. The daphnia should be washed thoroughly and fed to fish fry.

Collection: Daphnia are collected with the help of scoop net having 100–200-micron mesh.

Bloodworm: Chironomid larvae are commonly used as live food for maintenance of ornamental fishes.

Chironomid flies are attracted towards foul smell where they lay eggs which hatch into chironomid larvae.

The worm chironomid larvae are herbivorous in feeding habits and feed on algae, detritus, decaying organic and vegetable matter etc.

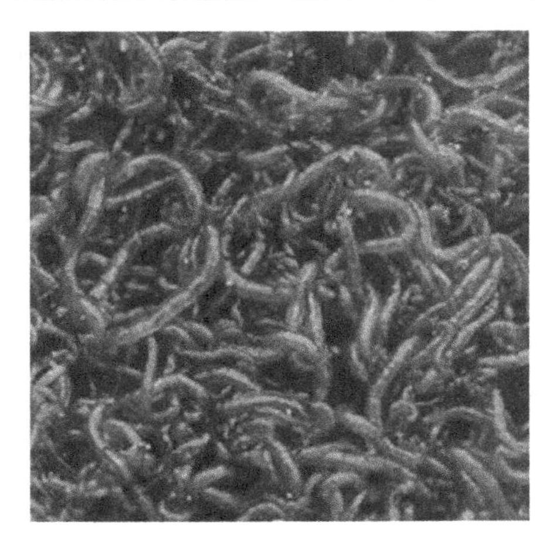

Culture

Chironomid can be cultured in 70 liters tank containing palm oil mill effluent (POME) and *Chlorella vulgaris* separately. The production of chironomid larvae is significantly higher in POME tanks (58 g. 20 liters POME) than chlorella culture (35 g / 20-liter chlorella).

Tubifex

It is good live food source and can be obtained all over the year. Their length varies from 3/8 inches to 2inches long. They are extremely thin, rusty red in colour and found plenty in mud, streams and ponds. They are collected by lifting the mud with a spade and putting it into muslin bags. Then swell the bags in streams remove all the worms. They form a ball like mass in shallow container. Worms are stored in a bucket or large bowl in to which a slow steady stream of water is allowed to enter.

Tubifex worms are long and slender, there is no distinct head. At the extreme anterior end there is a lip like structure known as the 'postomium', which is not counted as segment.

They are also abundantly found in places where suspended material from sewage settles in streams of reduced flow rate. These areas are known as "sludge banks".

Tubifex is one of the most reliable indicators of organic pollution. These worms live partly buried into the sediments.

Tubifex worms feed on decaying organic matter, detritus, vegetable matter which are commonly available in sewage drains.

Tubifex worm is hermaphrodite, because it has both male and female in the same animal.

Culture

Tubifex can be easily cultured on mass scale in containers with 50 to 70 mm thick pond mud at the bottom , blended with decaying vegetable matter and masses of bran and bread. Continuous mild water flow is to be maintained in the container, with a suitable drainage system. Within 15 days cluster of tubifex worm develop and this can be removed with mud mass.

These worm are considered as one of the best , cheap and most easily available live food which can be fed to fin fishes. They are cultured in commercial scale and sold to aquarist to feed graceful ornamental fishes and other fish culturist. The worms grow fast on a substance containing 75 percent cow dung and 25 percent fine sand with continuous running water at the rate of 250ml/min in a culturing system of 150×15×15 cm. Addition of cow dung at the rate 250mg/cm once in 4 days give optimum growth and the worms can be harvested at a rate of 125mg/cm/30 days. The worms can be harvested before dawn or after dusk since they are photophobic.

Microworms

Microworms are an ideal food for young livebearers because of their small size less than 1/8 inch (3mm) to avoid any unpleasant smell, and to keep the worms thriving, a new culture should be started every five to six days. Shallow plastic dishes make good containers and these should have a little pre-cooked cereal, made into a stiff paste, spread thinly over the bottom of the dish. The starter culture is added by smearing over the surface of the cooled food. It is covered with a sheet of glass and keep at room temperature. After a few days the surface of the culture will be found to be seeing with the tiny worms. If a knife –blade is scraped around the inner edge of the dish the worms will be collected; clean and ready to swill into the fish tank. After about five days the culture will start to “go off”. Fresh cultures will need to be started, preferably arranging a slight overlap in time between each culture.

Cyclops

Another small crustacean so named because of its single eye. They feed upon similar foods to the daphnia and are generally most abundant the warmer months of the year. Cyclops are found in the same waters as daphnia and will often be found as part of the daphnia haul. They are good food, especially for the smaller fish.

Mosquito Larvae

Mosquito larvae is referred to as wrigglers. Almost black in colour, they are up to ¼ inch (16 mm) long and head down from the water surface when disturbed they leave the surface with a whipping movement but, due to their dependence upon microscopic food. Mosquito larvae will be found in most still waters.

Culture

Cow dung is the medium for culture of mosquito larvae. They can be harvested using scoop net.

Artificial Feeds

The artificial feeds are broadly classified into dried, freeze dried and frozen diets

Dried foods such as flakes, pellets, sticks and tablets are widely available mostly dried foods have a limited shelf life and should be used

within a few months of purchase. Although they can still be used after this period, the vitamin levels decrease with time.

Freeze-dried and frozen fish diets included mosquito larvae, bloodworms, daphnia and brine shrimp.

Nutrients

Proteins (amino acids), lipids (fats, oils, fatty acids), carbohydrates (sugars, starch), vitamins and minerals are important nutrients required for good growth and health of ornamental fishes. In addition, pigments (carotenoids) are commonly added to the diet of ornamental fishes to enhance their flesh and skin colouration.

Proteins and Amino Acids

Fish meal, soybean meal, fish hydrolysate, skim milk powder, legumes, and wheat gluten are excellent sources of protein, additionally the building blocks of proteins (free amino acids) such as lysine and methionine are commercially available to supplement the diet.

Lipids

Oils from marine fish, such as sardines, and vegetable oils from canola, sunflower and linseed, are common sources of lipids in fish feeds.

Carbohydrates

Cooked carbohydrates, from flours of corn, wheat or other "breakfast" cereals, are relatively inexpensive sources of energy that may spare protein (which is more expensive from being used as an energy source).

Vitamins and Minerals

The variety and amount of vitamins and minerals are so complex that they are usually prepared synthetically and are available commercially as a balanced and pre-measured mixture known as a vitamin or mineral premix. This premix is added to the diet in generous amounts to ensure that adequate levels of vitamins and minerals are supplied to meet dietary requirements.

Pigments

A variety of natural and synthetic pigments or carotenoids are available to enhance colouration in the skin of freshwater and marine ornamental fish. The synthetically produced pigment, astaxanthin is the most commonly used additive (100-400 mg/kg). Cyanobacteria (blue-green algae such as spirulina), dried shrimp meal, shrimp and palm oils, and extracts from marigold, red peppers are excellent natural sources of pigments.

Binding Agents

Another important ingredient in fish diets is a binding agent to provide stability to the pellet and reduce leaching of nutrients into the water. Beef heart has traditionally been used both as a source of protein and as an effective binder farm-made feeds. Carbohydrates (starch, cellulose and pectin) and various other polysaccharides, such as extracts or derivatives from animals (gelatin), plants (gum Arabic and locust bean), and seaweeds (agar, carageenin, and other alginates) are also popular binding agents.

Preservatives

Preservatives, such as antimicrobials and antioxidants, are often added to extend the shelf-life of fish diets and reduce the rancidity of the fats.

Vitamin E is an effective, but expensive, antioxidant that can be used in laboratory prepared formulations. Commonly available commercial antioxidants are butylated hydroxyanisole (BHA), or butylated hydroxytoluene (BHT), and ethoxyquin. BHA and BHT are added at 0.005 per cent of dry weight of the diet. While ethoxyquin is added at 150mg/kg of the diet. Sodium and potassium salts of propionic, benzoic or sorbic acids, are commonly available antimicrobials added at less that 0.1 per cent in the manufacture of fish feeds.

Attractants

Other common additives incorporated in the fish feeds are chemo attractants and flavorings, such as fish hydrolysates and condensed fish solubles (typically added at 5 per cent of the diet). The amino acids glycine and alanine, and the chemical betaine are also known to stimulate strong feeding behavior in fish. Basically, attractants enhance feed palatability and its intake.

Formulation and Preparation of Artificial Feed

Nutritional Requirements

Ornamental fishes require protein, fat, carbohydrate, vitamin and minerals as like other vertebrates. Protein helps to buildup muscle and tissue. Fat and carbohydrate provides energy. Vitamin and mineral mixtures are necessary to make the fishes resistant to disease and strengthen their bones. Apart from these, binders like guar gum or carboxy methyl cellulose is added at 0.2 per cent to 1 per cent level to increase the water stability of feed. Pigments like carotenoids are added to improve the colour of the fish.

	Protein	Fat	Carbohydrate
Small fishes	40-45 per cent	4-6 per cent	40 per cent
Adult fishes	30-35 percent	6-8 per cent	50 per cent

Artificial feed is classified into three major categories according to the preparatory methods.

Dry feed: The feed is in dried form with 8-10 per cent moisture. It is further classified into 5 sub-groups.

- Pellets
- Flakes-are flat in structure.
- Freeze dried feed- they are in frozen form
- Tablet form
- Granular form-they are very small and round shape. They are similar to grains

Moist feed: It is prepared daily and fed to fishes- the moisture content is 35 per cent.

Paste feed: This is mainly prepared for young ones of all. The feed ingredients are made into paste and fed to fishes squeezing through mesh.

Feed formulation: Since protein plays an significant role on the growth of fishes, protein level is given importance while formulating the feed. To balance and formulate the required protein diet, Pearson square method is followed.

Feed formulation of livebearers:

Guppy (protein-15-34 per cent)

Ingredients	(per cent)
Wheat bran	70-80
Fish meal / shrimp meal	10-15
Soybean meal/skimmed milk/butter milk powder	15-20

- Supplemented with tubifex for adult.
- Supplemented with moina for fry.

Platy and swordtail (18-21 per cent protein)

Ingredient	Per cent
Fish meal	60-80
Wheat bran/flour	20-40

Molly (12 per cent protein)

- Bread crusts
- Supplemented with phytoplankton + zooplankton + algae

Feeding Strategies

- "Where", "When" and "How" to feed the fish should be taken prior considerations. Some fishes are mid-water and some are bottom feeders many learn new habits in captivity but some don't, especially if the mouth position dictates feeding posture.
- Feed size is equally vital. It should be small enough to be ingested easily by the fishes.
- Optimal level of feeding should be carried out overfeeding and underfeeding should be avoided.
- Overfeeding pollutes the water quality due to the accumulation of uneaten feed and fecal matter.
- Under feeding results in reduced growth rate in fishes.
- The amount of feed given per day is calculated as follows
- Amount of feed given/per day = biomass × percentage of body weight of feed per day.
- Biomass = stocking density × survival rate × average body weight
- Feeding two (or) three times a day with sufficient food as they will consume in a few minutes is good to be provided.

CHAPTER 7 Water Quality Management

"Water is the Driving Force of the Nature"

Water is the substantial element for all life forms. Especially, in fishes the water is the environmental and the metabolic medium in which they survive. The fishes pass water through their alimentary canal and excretory system and also "breathe" it using their gills to extract the oxygen and carbon dioxide.

Water is highly variable in its chemical properties; it can be hard or soft; fresh, brackish or saline; acidic or alkaline; rich or poor in oxygen; pure or polluted. The fish species preference varies in the water chemistry. For instance: guppy prefers soft water for the growth wherein mollies and platies prefer hard water. Therefore, we must not assume that water is common element for all varieties of fish and they are happy with whatever that comes out from the tap.

The first step to be a proper aquarist is to be a "water technician", knowing the appropriate environmental conditions required by the fish species to be cultured and the best management of the water quality.

In the case of ornamental fish farming the main source of water is obtained from the rivers, lakes, canal, reservoirs and artesian wells. The major physico-chemical characteristic features to be considered for the ornamental farming are temperature, pH, hardness, dissolved oxygen, carbon dioxide and chlorine.

Temperature

Water temperature is the most prime variable influencing the growth, breeding and transportation of the ornamental fishes. The fishes are poikilothermic or "cold blooded", that means their body temperature is nearly similar as the temperature of the water surrounding them. Thus, any changes in the water temperature directly affects the body temperature of the fishes. The rate of all the biochemical processes is temperature dependent. Within the temperature range that normally occurs in the natural habitat of a particular species the rate of biochemical processes roughly doubles for every 10°C increase in temperature. Majority of the tropical ornamental fishes can tolerate water temperature between 21°C to 30°C, while livebearers have a good growth rate at 22°C to 25°C. The most preferable temperature for the breeding of the livebearers is 27°C. In order to maintain optimal temperature range, suitable devices can be used such as water heaters during the winter season. The swordtails are very sensitive to 25°C to 27°C. Water temperature is a vital factor affecting the dissolved oxygen budget in the culture system and low dissolved oxygen are encountered mostly at high temperatures. The water temperature also a play an indirect role in the health of the aquatic animals because it strongly influences the occurrence and outcome of the infectious diseases.

pH

The pH value expresses the intensity of the acidic or basic nature of the water. It is defined as the negative logarithm of the hydrogen ion activity which, in fresh water, is essentially equal to the hydrogen ion concentration. The conditions become more acidic as the pH value decreases and more basic as the pH value increases. The fishes exposed to extremes of pH can be stressful or lethal, but the indirect effects of pH and interactions of pH with other variables are usually more vital in ornamental farming than the direct toxic effects. The ornamental livebearers are hardy species and can tolerate a wide range of pH, mostly they prefer neutral to alkaline pH ranging from 6.5-8.5. If the pH of the water falls below the desired level the pH can be increased by liming the water. The pH has a diurnal fluctuation pattern which is associated with the intensity of photosynthesis. This is because carbon dioxide is required for photosynthesis and accumulates through night time respiration. It peaks before dawn and is at its minimum when photosynthesis is intense. All organisms respire and produce Carbon dioxide (CO_2) continuously, so that the rate of CO_2 production depends on the density of organisms. The rate of CO_2 consumption depends on phytoplankton density. Carbon dioxide is acidic and it decreases the pH of water. Also, at lower pH, CO_2 becomes the dominant form of carbon and the quantity of bicarbonate and carbonate would decrease. The consumption of CO_2 during photosynthesis causes pH to peak in the afternoon and the accumulation of CO_2 during dark causes pH to be at its minimum before dawn. Therefore, the pH should be monitored before dawn for the low level and in the afternoon for the high level. The magnitude of diurnal fluctuation is dependent upon the density of organisms producing and consuming CO_2 and on the buffering capacity of pond water (greater buffer capacity at higher alkalinity). The control of pH is vital for minimizing ammonia and H_2S toxicity.

Hardness

Total hardness is the sum of concentrations of calcium and magnesium in water, expressed as mg/L equivalent to CaCO3. The total hardness helps in classifying the waters into 'soft water' and 'hard water'. The optimal range of hardness for normal growth of the ornamental fishes is 100-300 ppm. The water with less than 5ppm hardness affects the growth of the fishes and eventually leads to death of the fish. Guppies prefer soft water for better growth, colouration and production. Whereas, the mollies and platies prefer hard waters.

Dissolved Oxygen

Oxygen is one environmental parameter that exerts a tremendous effect on growth and production through its direct effect on feed consumption and metabolism and its indirect effect on environmental conditions. Oxygen affects the solubility and availability of many nutrients. Low levels of dissolved oxygen can cause changes in oxidation state of substances from the oxidized to the reduced form. Lack of dissolved oxygen can be directly harmful to culture organisms or cause a substantial increase in the level of toxic metabolites. It is therefore important to continuously maintain dissolved oxygen at optimum levels of above 3.5 ppm.

There is above 21% oxygen content in the air. Air acts a big reservoir for oxygen concentration, in water it is limited due to its solubility. The solubility of oxygen:

- decreases as the temperature increases.
- decreases exponentially with increase in salinity.
- decreases with lower atmospheric pressure and higher humidity
- increases with depth.

Photosynthesis plays a major role in oxygen production; respiration of all living organisms in the pond. It is the major factor involved in oxygen consumption. Oxygen concentration in pond water exhibits a diurnal pattern, with the maximum occurring during the peak of photosynthesis in the afternoon and the minimum occurring at dawn due to night time respiration. Atmospheric oxygen crosses the air-water boundary and dissolves in the water matrix. The only way that oxygen can be introduced from air to water is by diffusion. Atmosphere contains vast amount of oxygen, some of which diffuse into pond waters when they are unsaturated with oxygen. Likewise, oxygen is lost to the atmosphere when pond water has supersaturated with oxygen. The driving force causing net transfer of oxygen between air and water is the difference in the tension between oxygen in the atmosphere and oxygen in the water. Once equilibrium is reached i.e., oxygen tensions in air and water are the same, the net oxygen transfer ceases. Oxygen must enter or leave a body of water at the air-water interface and for the very thin film of water in contact with air, the greater the deficit or surplus, the faster oxygen will enter or leave the film. In general, the rate of diffusion of oxygen depends primarily on the oxygen deficit in water, the amount of water surface exposed to the air and the degree of turbulence. Typically, dissolved oxygen is measured either in mg. per litre (mg/l) or parts per million (ppm) with O ppm representing total oxygen depletion and 15 ppm representing the maximum or saturation concentration. In aquarium tanks fishes will undergo stress and will be liable to parasitic attack if the optimal oxygen levels are not maintained. The oxygen level in the aquarium tanks can be enhanced by constant aeration, sprinkling of water, surface agitation, etc. the surface agitation of water apart from enhancing the oxygen level, drives out free carbon dioxide form the water.

Carbon-dioxide

Carbon-dioxide is highly water-soluble, biologically active gas. It is produced during respiration and consumed in the process of photosynthesis. Thus, the concentration of dissolved carbon dioxide usually varies inversely with dissolved oxygen. Dissolved carbon-dioxide should be taken care because:

- It can be a stressor to the fishes.

- It influences the pH of the waters.
- It is a nutrient required for the plant growth and its availability may limit the primary productivity of some aquatic ecosystems.

High concentrations of dissolved carbon-dioxide (more than 60-80 mg/l) have a narcotic effect on fishes and even higher concentrations may cause death. In ornamental farming the free carbon-dioxide at a concentration of more than 15ppm is detrimental to the ornamental fishes.

Chlorine

Chlorine may sometimes be present in municipal water that is used for holding fish or it may be applied to holding facilities or ponds to effect disinfection. All chlorine residuals are oxidizing agents, although oxidizing strength ability they can induce lesions which varies among the different forms. The primary mode of toxigenesis by chlorine residuals is oxidation of gill membrane tissues. Irritation and cellular damage result in generalized gill lesions, including excessive mucus production, hyperplasia, inflammation, hypertrophy and necrosis (Heath 1987; Mitchell and Cech 1983). The growth and survival of any sensitive fish are affected by chlorinated tap waters. Chlorine content as low as 0.1 ppm itself is toxic to the fishes. The chlorine content in the water can be eliminated by heating the water, squeezing the free end of the hose firmly, so that, the water flows out in a broad stream. This chlorine-free water can be used in the ornamental tanks. Alternatively, water containing the chlorine should be kept over-night where the chlorine escapes and the water becomes usable.

Ammonia

In aquaria and ponds the principal sources of ammonia are:

- Excretion by fish and other livestock as a normal part of their metabolism

- The breakdown of protein in uneaten food or dead livestock that remains undetected.

It is therefore of great importance that careful cleaning is undertaken at suitable intervals. As ammonia is released into the water by either of these processes it may take one of two forms:

- Free Ammonia (unionised ammonia, NH_3). This form of ammonia is highly toxic to fish
- Ammonium (ionised ammonia, NH_4+). This form of ammonia is virtually nontoxic to fish

The balance between free ammonia and ammonium is determined by the pH and temperature of the water. OATA recommends that free ammonia should not exceed 0.02mg/l in freshwater. The ammonia levels can be reduced by:

- Dilution by water changes.
- Reduction of stocking densities, improvement of feeding and general husbandry procedures.
- Improvement of biological filtration.
- Use of ion exchange materials.

Nitrite

In aquaria and ponds nitrites are produced by bacteria (including *Nitrosomonas* spp.) when ammonia is broken down. OATA recommends that Nitrite should not exceed 0.2mg/l in freshwater. Nitrite poisons fish by binding the haemoglobin in the blood preventing it from carrying oxygen, in effect suffocating the fish. The gills of fish dying as a result of nitrite poisoning are a characteristic brown colour. In freshwater the toxicity of nitrite may be reduced by the addition of small amounts of certain chloride salts. The other ways to reduce the nitrite are:

- Reduction of stocking densities, improvement of feeding and general husbandry procedures.
- Improvement of biological filtration.
- Dilution by water exchange.

Nitrate

Nitrates are:

Produced as bacteria (including *Nitrobacter* spp.) which breaks down nitrites;

- Introduced in tap water. In some areas of the country tap water nitrate levels exceed 130mg/l. Nitrate is generally of low toxicity though some species, especially marine, are sensitive to its presence. When nitrate levels are high, as a result of biological filtration, other chemicals produced in this process may be present at levels that adversely affect fish health. OATA recommends that nitrate levels in freshwater systems do not exceed those in the tap water supply by more than 50mg/l. The nitrate levels can be reduced by
- Water exchange
- Use of ion exchange material
- Increase plant density
- Use of denitrifying biological filtration.

Water quality parameters	Optimum level
Temperature	22°C-25°C
pH	6.5-8.5
Hardness	100-300ppm
Dissolved oxygen	>3.5ppm
Ammonia	< 0.02mg/l
Nitrite	< 0.2mg/l
Nitrate	< 50mg/l

Water quality parameters required for live-bearing fishes

Water Quality Management

Water quality parameters should be monitored to serve as guide for managing the ornamental fish culture system so that conditions that can adversely affect the growth of fishes can be avoided. In cases where problems are encountered, these parameters can help in the diagnosis, so a remedy can be formulated. Individual parameters usually do not tell much, but several parameters put together can serve as indicators of dynamic processes occurring in the system. Careful monitoring and data collection will remain useless unless it influences decisions regarding water management. This becomes more important as cost to implement various management schemes (aeration, water exchange, inputs) increase. Most of the water quality problems can be solved with adequate water exchange. Thus, if large quantities of water suitable for ornamental fish culture were available, monitoring would not be as critical and high production levels can be targeted. If water is limited, the risk of encountering water quality and disease problems increases as one goes for more intensive culture.

Water Exchange

Water exchange is the most effective and widely employed method to maintain good water quality besides water quality enhancers like sanitizers, zeolite etc. Generally, water exchange is used to adjust salinity as desired, to remove excess metabolites, to keep algae healthy to produce ample oxygen and to regulate pond water temperature. The exchange rate varies with the production period, stocking density, total biomass, levels of natural productivity, turbidity, source and volume of water.

The principle of water exchange is to change the water in a way such that the water quality changes gradually instead of abruptly. In semi-intensive systems, frequent and sometimes even continuous water exchange at a small flowing rate is employed. Abrupt addition of large quantities of water in small ponds may result in sudden environment change, which subsequently can cause stress in culture organism. Therefore, massive water replacement is not recommended unless there is sudden die-off of plankton, critical low oxygen or after the application of chemicals. Lowering the water level first and then adding new water is not recommended, especially during daytime in summer. Increasing water temperature, while lowering water level, can reduce the capability of water to hold oxygen and hasten the degeneration of the pond bottom, leading to oxygen depletion. It is better to add new water first according to the predetermined exchange rate.

Aeration

The aerators are used to increase contact surface of water with air thereby increasing the area through which oxygen is absorbed by the water and to create a circular movement of the pond water. This has the following advantages:

- It increases the dissolved oxygen level of the water and prevents oxygen depletion during the night.
- It accelerates the diffusion effect of not only the oxygen, but also enables the capture or release of carbon dioxide. Carbon dioxide is important for culture of algae and therefore for maintenance of appropriate water colour.
- It facilitates the volatilization of undesirable gases such as N_2, NH_3, CH_4 and H_2S · It reduces the daily fluctuation range of pH value.
- It accelerates the decomposition and mineralization of organic matter in water and soil and helps in the release of nutritive value of fertilizers.
- It diminishes the possible stratification of pH, DO, salinity and temperature in the pond water.

- It helps in mixing the pond water and maintenance of ideal conditions all over the pond.
- It increases turbidity when necessary.

Phytoplankton Management

Phytoplankton play a significant role in stabilizing the whole pond ecosystem and in minimizing the fluctuations of water quality. A suitable phytoplankton population enriches the system with oxygen through photosynthesis during day light hours and lowers the levels of CO_2, NH_3, NO_2 and H_2S. A healthy phytoplankton bloom can reduce toxic substances since phytoplankton can consume NH_4 and tie-up heavy metals. It can prevent the development of filamentous algae since phytoplankton can block light from reaching the bottom. A healthy bloom also provides proper turbidity and subsequently stabilizes fishes and reduces cannibalistic behaviour of the parental fishes. It decreases temperature loss in winter and stabilizes water temperature.

Removal of Dissolved Metabolic Organics

One of the important stress factor is the increase of dissolved metabolic organics in culture water. It can increase ammonia and microorganisms leading to water quality deterioration and causes high mortality rate. To prevent the build-up of dissolved organics, frequent partial to total water exchange is necessary; or the pollution could be reduced by the chemically removing the pollutants by adsorption using activated carbon. The best way to facilitate the removal of metabolic wastes in a pond is by flushing out water from the bottom. Constantly maintaining high DO in the pond through supplemental aeration and water exchange, enhances nitrification. Nitrification is a major mechanism for ammonia removal in well-aerated ponds.

Pond Bottom Treatment

The pond bottom should be completely dried and aerated to get rid of toxic gases. Many ponds in low-lying areas cannot be completely drained and dried. To overcome this, the farmers apply waste digesters to the ponds. The digesters are harmless bacteria (probiotics) and enzymes that consume organic matter on the pond bottom. After the application of digesters farmers apply a disinfectant, either organic silver or organic iodine. Organic silver is highly effective against bacteria and viruses and its toxicity to aquatic life is very low. Organic silver is applied at the rate of 18 litres per hectare after lowering the water depth to 12 inches. Seven days after the application, this disinfectant disintegrates, so there is no need to flush the pond. Organic silver also prevents the development of algae that grows on shells. Organic iodine, can cure gill or shell diseases, kills bacteria on contact and has low toxicity. Its effect can be noticed within 24 hours and the pond bottom can be disinfected without emptying the pond. The suggested dosage is 5 ppm to 10 ppm. Its affectivity lasts for two to three days compared to about seven days in the case of organic silver.

Self-pollution as a Possible Factor

When the accumulation of nutrients within ponds is high, self-pollution of the culture environment reduces production, frequently as a result of a severe disease-outbreaks. Although, in some cases, production losses can be linked back directly to disease outbreaks, it is often difficult to separate the effect of disease and poor water quality. Disease-outbreaks occur when

(1) a pathogen infects a population previously not exposed to the microorganism.

(2) poor culture conditions weaken resistance to pathogens permanently present in the culture environment.

Outbreaks of new infectious diseases will be difficult to prevent as long as there are no strict regulations for transfer of culture stocks between regions. From a practical point of view, more attention should be paid to culture conditions, with special attention to water quality.

CHAPTER 8

Disease and Health Management

Disease and health management in an ornamental system involves three important components, the host, pathogens and the environment. For a disease situation to exist, there should be a potential pathogen, a suspectable host and environment condition that bring about either increased virulence of the pathogen (or) decreased resistance of the host. The disease in the aquaria is generally classified into two categories:

- Non infectious diseases
- Infectious disease

Non Infectious Disease

They are non-communicable disease. It arises mainly due to altered environmental parameters which causes stress to the fish and affects its immune system.

DEPLETION OF DISSOLVED OXYGEN (DO)

Fish, feed waste, microbe and plants during night consume DO. Any change in the input particularly density/crowding and waste accumulation cause DO depletion.

Clinical signs: surfacing, gulping with mouth wide open

Remedy: Aeration, waste removal and filtration.

EXCESS OXYGEN

Causes bubble trap in blood vessels, on gills and body leading to discomfort and deaths.

Remedy: Reduce plants and light

ACID OR ALKALINE WATER

pH less than 5and more than 9 are not desirable. pH depends on the sources of water.

Clinical signs: milky discoloration of skin due to acidosis. Ragged conditions of fins and irritation to gills due to high pH.

Remedy: Neutralize acid water or change the water source. To reduce pH, reduce the plants as more plants produce more CO and dicarbonate.

TEMPERATURE

Ideal temperature for warm water fish is 20-28°C. Low temperature stresses the fish resulting in reduced immunity and increased infection;

ACCUMULATION OF NITROGEN WASTE

Accumulation of organic matter and fish excreta are sources of ammonia, nitrite and nitrate. Ammonia and nitrite are toxic and not desirable even at sub lethal level. Accumulation of nitrogenous waste is common in new tanks.

Signs: gills and skin damage, pale color and prone to disease.

Remedy: Establishing good bio-filters, and plants for using nitrate.

CHLORINE

Chlorine sources in municipal water or residues of disinfection in aquaria.

Signs: pale gills with white edge.

Remedy: Tap water has chlorine from 0.2 to 0.5ppm. store water for 48h for dechlorination or pass the water through activated charcoal or use sodium thiosulphate.

EXOGENOUS MATERIALS

- Lead, zinc copper can enter aquarium through materials used in pipe line and other materials in tank construction.
- Chemicals used for treatment.
- Heavy metals from water supply, food or leached from equipment.

NUTRITIONAL IMBALANCE

Poor health due to incorrect nutrition is a common. A balance of artificial and natural feed is necessary. Fattening of fish is common problem due to luxury feeding and less activity of fish in aquaria Egg binding due to inability of aquarium fish to reabsorb egg as in normal condition is common leading to health hazard. Intestinal inflammation is another common ailment due to imbalanced feeding.

MISCELLANEOUS FACTORS

- Vandalism: many ornamental fishes suffer severe damage at the hands of brick throwing vandals.
- Damage due to excessive handling or violent spawning behaviour.
- Overstocking can be extremely stressful to the fishes.

INFECTIOUS DISEASE

The communicable diseases are known as infectious diseases. Most of the diseases occurring in live bearers are parasitic and bacterial diseases. The sign of diseases is found on the external surface.

PROTOZOAN DISEASES

Ich Disease (Ichthyphthiriosis)

It is common disease in fresh water, brackish water and marine aquaria caused by ichthypthirius, a protozoan ciliate. It is a skin parasite.

Signs: Salt-like specks on the body/fins, clamped fins, excessive slime.

Remedy: Quarantine for 1 week, Acriflavine 1 g/100 l, quinine 1g/ 100 l for 9-10 days, methylene blue 1g/100 l

Velvet Disease

Also called "blue slime disease" is caused by costia species, a protozoan disease, feeding on skin and gill

Signs: folded fins, weak, rubbing body against sides and bottom of the tank

Remedy: acriflavine 1 g/100 1 for 2-3 days.

Costia:

Signs: Milky cloudiness on the skin

Remedy: acriflavin 1 g/100 1 for 2-3 days.

PARASITIC DISEASES

Gyrodactylus:

They are common on gills.

Remedy: 10g common salt/litres for 20 min

Argulus (fish louse):

Argulus is a flattened mite-like crustacean which attaches itself to the body of fish.

Signs: Retracted fin, scratching movement.

Remedy: Bath treatment with potassium permanganate (10mg/l) can be given. In case of large fishes, the lice can be picked off using forceps.

Ergasilus:

It is a fresh water parasite found on the gills.

Signs: pale gill, emaciation.

Remedy: lindane can be given.

Lernaea:

Signs: scrapping movement, whitish-green threads hang out of the fish skin.

Remedy: Bath treatment with potassium permanganate (10mg/l) can be given.

Worms:

Infection with worms such as round worms and acanthocephalan is common in aquarium fish. It affects general wellbeing of fish.

BACTERIAL DISEASES

Dropsy:

It is caused by *Aeromonas hydrophila*, *Pseudomonas* and opportunistic pathogens.

Signs: Scale protrusion, accumulation of ascites body fluid, popped eye and bloating of the body.

Remedy: Antibiotics such as chloromycetin and tetracycline at the rate of 10mg per litre is effective.

Fin Rot and Tail Rot

Fin rot and tail rot appears to be a bacterial infection of tail and fin caused generally by poor conditions or fin nipping partners.

Signs: Margin of fin affected to begin with followed by inflammation, reddened areas at base of fins and sometimes loss of whole fin.

Remedy: Antibiotics.

Red pest:

Signs: bloody streaks on the body.

Remedy: Acriflavine (0.2 percent) should be used as disinfectant at rate of 1ml per litre.

Mouth fungus:

It is caused by the bacterium *Flexibacter columnaris.*

Signs: white cotton patches around the mouth, grey or white lines around the lips.

Remedy: Penicillin at 10,000 units per liter is a very effective treatment. Second dose with chloromycetin, 10 to 20 mg per liter may be given after two days.

FUNGAL DISEASES

Saprolegnia:

The fungus cotton-like growth on the skin, covers large areas of fish. The fish egg turns white. It also leads to secondary infections.

Signs: Greyish or whitish growth in the skin and fins of the fish,

Remedy: Malachite green 0.2g/1 for short duration.

Common diseases of live-bearing fishes and treatment methods:

Common disease	Name of the fish	Treatment
White spot disease	Molly	Dip treatment in 5% methylene blue for 2 days or addition of 2% mercurochrome solution in the infected tank
Fungal disease	Guppy, molly and sword tail	Bath treatment in 0.1 ppm malachite green for 7 days
Bacterial diseases (fin rot, red spot, tail rot)	Molly, platy, swordtail and guppy	1 minute dip treatment in 500 ppm copper sulphate solution or bath trypaflavine or 0.005% chloromycetin or sulfanilamide
Dropsy	Molly, guppy and swordtail	Sulfadiazine injection at 7mg/ 100g or bath treatment in 0.001% chloromycetin or dip treatment for one minute in 1: 2000 copper sulphate solution for 3-4 days
Ectoparasites	Guppy, molly, platy and sword tail	0.1% potassium permanganate or removal by forceps or 4% common salt solution for 10 to 30 minutes
Ich (Ichthy-phthiriosis)	Guppy, molly, platy and sword tail	Quinine hydrochloride at 30mg per litre

Health Management in Livebearers

"Prevention is better than cure"

The shipshape health management of the livebearers is based upon the concept of prevention. Prevention of disease is always cheaper than dealing with a disease outbreak once it has occurred. Disease prevention is a cornerstone factor when developing a sustainable ornamental fish industry. There are ten steps which helps in quality health management of the live bearers. The ten steps are:

- Define your goals
- Know your fish
- Understand the water quality
- Understand biosecurity
- Plan for healthy animals
- Identify reliable resources
- Practice regular health management
- Recognize the disease
- Work the problem
- Re-evaluate continuously\

1. DEFINE your goals

- Formulation of realistic health management goals is prerequisite for the successful operation of the facility.
- Setting these goals must be based upon your level of commitment, finances, facility size and infrastructure, staff numbers, level of staff training and diseases of concern.
- Some of the questions to consider when defining the fish health management goals include:
    - Do you want to maintain/sell healthy fish?
    - What species do you wish to culture? Do they have any specific husbandry requirements?
    - Do you have the ability to provide for these specific husbandry requirements?
    - Who will be your suppliers? Can they provide your facility with healthy fish?
    - Who will be your customers? What are their expectations regarding animal health?

- What is you animal holding capacity? Is this capacity large enough to meet the animal's needs as well as your business needs?
- Are you willing to invest the resources to develop a reasonable fish health management program for your facility and to create a bio secure facility?
- Must your facility adhere to any international, federal or state animal health and welfare regulations and guidelines?
- Do you have the capability to meet these regulations and guidelines?
- What are the common diseases of the species you wish to culture?
- Are you willing to invest in training for your staff about the importance of daily health management and biosecurity?
- Are you willing to invest in continuing education for you and your staff as knowledge and techniques evolve?
- Are you willing to implement and maintain a quality control program?

- Spending time to think about your answers to these key questions will be invaluable as you begin to develop an effective fish health management program tailored to your facility.

2. Know your fish

- It is imperative that facility managers should have a clear idea about the basic natural history, biology and husbandry requirements of the fish under their care.
- They should know about the following details : country and environment of origin, life cycle, maximum size, behaviour in the nature and captivity, temperature and required water quality parameters, key breeding requirements, optimal captive husbandry requirements (tank/pond size, habitat needs, optimal stocking

density, and life-support requirements needed to maintain optimal environmental quality), common diet and feeding frequency, common signs of health and disease, common diseases and treatments.

3. Understand the water quality

- Water quality parameters are the vital factors leading to a disease outbreak among captive ornamental fish.
- It is very important to know the key water quality parameters to measure in a particular system, know how and when to measure them properly, understand how to interpret those readings in terms of their effect on fish health as well as on the biological life support (nitrification and the carbonate cycle) within the holding system.

4. Understand Biosecurity

- Biosecurity consists of the practices and procedures used to prevent the introduction, emergence, spread, and persistence of infectious agents and disease within and around fish production and holding facilities.
- It is the cornerstone for successful fish health management.
- A typical biosecurity program focuses on two major components:
  - Pathogen exclusion
  - Pathogen Containment
- Pathogen exclusion is the procedures implemented to prevent pathogen introduction to a facility. These procedures typically focus on reducing the risks associated with various potential routes of disease entry into a facility and establishing an effective quarantine protocol for all new introductions.

- The most important pathways for pathogen entry are associated with:
    - Fish
    - Water
    - Food
    - People
    - Equipment
- Quarantine: Quarantine is the final critical component of pathogen exclusion for your facility. The major role of quarantine is to prevent the introduction of pathogens directly into your retail population by separating new arrivals.
- Quarantine also provides for the important process of acclimatization of fish to new water conditions, new husbandry protocols, and new feeds.
- Furthermore, the quarantine system and quarantine period allows time for the fish immune system to recuperate from the stresses of transport and handling.
- Pathogen Containment is containment that ensures a disease outbreak in an ornamental fish facility is confined to the smallest possible area.
- Pathogen containment requires understanding of three key aspects of disease:
    1) The factors that influence disease resistance/ persistence within the fish as well as the aquatic environment.
    2) The locations in which pathogens may persist within a facility. These locations include: the fish, invertebrates, plants, water (even the source water), the life support system, system surfaces, and equipment used during regular husbandry procedures.

3) The means by which pathogens may be transmitted to susceptible fish species. These routes of transmission include: waterborne, airborne, vector borne (organisms that may harbor or transmit pathogens from one fish to another), fomite transmission (inanimate objects on which pathogens can be transmitted from location to location) and food borne.

An understanding of these key factors is essential to implementing an effective biosecurity program that ensures critical areas of risk are not overlooked

5. Plan for healthy animals

- Planning for healthy animals typically begins with an examination of the animal facilities from the standpoint of layout and workflow, with an eye towards minimizing stress in the animals and risks of disease introduction and spread within the facility.
- Any system that is not designed for ease of maintenance will typically not be maintained at the optimum level. Similarly, husbandry protocols developed without a thorough understanding of the actual husbandry procedures will not be easily followed.
- A strong quality management program will help make these planning efforts successful. Detailed standard operating procedures (SOP's), regular record keeping, and ongoing staff training will help guide and reinforce all of the policies and procedures.

6. Identify reliable resources

- It is necessary to critically evaluate all aspects of the resources utilized in your ornamental fish operation.
- Ensure to procure the best quality fish for brood stock, from local sale or export.

- Be sure that you have the highest quality, nutritionally complete food supply you can afford for your livestock.
- Ensure that the equipment you purchase for your operation is reliable, tested, and calibrated. Ensure that you have adequate disease diagnostic laboratories available that can provide a timely and proper disease diagnosis and treatment recommendations.
- Finally, critically evaluate the information you use to make decisions about fish biology, husbandry and health management.

7. Practice daily health management

- Quality health management relies upon all staff adhering to their training and the standard operating procedures for that facility.
- It is compulsory that all staff, including the entry-level staff, understands the importance of SOP's, has been trained for each SOP, and knows the risks associated with failure to follow these protocols.
- Without this understanding we cannot guarantee the procedures will be consistently practiced.

8. Recognize the disease

- Learning to identify the diseased fish is key to recognizing if there is any health problem pertaining in the system.
- The earlier that compromised fish can be recognized, the greater the chance of averting a severe fish disease outbreak.
- It is important for all husbandry staff to become familiar with normal and abnormal behaviour and appearances for the fish species under their care.

9. Work the problem

The steps to be taken to work out the problem are:

- Address any immediate life-threatening problems first such as lack of power, air, and water flow.
- Next make a proper disease diagnosis.
- Finally, ensure that you have identified all of the potential contributing factors that may be associated with the disease outbreak.
- Recognize that many fish disease problems are multifactorial and it is very important to identify all contributing factors to ensure each is addressed by the treatment and management plan.
- Realize that many infectious disease problems are often precipitated by poor water quality, poor environmental conditions, and poor husbandry. If these factors are not corrected, even with a proper diagnosis and treatment, the condition may well return because the underlying causes have not been eliminated or controlled.

10. RE-EVALUATE continuously

- Continuously re-evaluating is the key element for the refinement of the health management strategy
- We have to consistently evaluate the status and health of the fish as well as the fish holding systems including pond/tank environment, water quality, food quality, aeration, filtration, life support, and back-up systems.
- We should realize that your SOP's are never set in stone and should be periodically reviewed and revised when necessary, including the husbandry and maintenance protocols, transport and acclimation protocols, and the quarantine and treatment protocols.

Conclusion

Finally, it is imperative that husbandry staff always rechecks the fish after any medical treatment to ensure that the treatment has been effective and the disease agent is under control and/or eliminated. Approaching the development and implementation of a health management plan using "The 10 Steps to Quality Fish Health Management" will not guarantee a disease-free facility, but will go a long way towards ensuring you have significantly reduced your risks of disease introduction and spread. Using a systematic approach to the key principles of fish health management will certainly aid you in the production of higher quality, healthier fish for your customers.

CHAPTER 9 Biosecurity in Ornamental Fisheries

Introduction

Biosecurity is a strategic and integrated approach that encompasses the policy and regulatory frameworks (including instruments and activities) for analyzing and managing relevant risk to human, animals and plant life and health, and associated risks to the environment. Biosecurity covers food safety, zoonoses, the introduction and release of living modified organisms (LMOs) and their products (e.g. genetically modified organisms or GMOs), and the introduction and management of invasive alien species. Thus biosecurity is a holistic concept of direct relevance to the sustainability of agriculture, and wide-ranging aspects of public health and protection of the environment, including biological diversity.

The overarching goal of biosecurity is to prevent, control and/or manage risks to life and health as appropriate to the particular biosecurity sector. In doing so, biosecurity is an essential element of sustainable agricultural development. (FAO,2006)

Biosecurity in Ornamental Fisheries

Biosecurity is the preventive measures taken against any infectious disease outbreak in the ornamental realm. It is the amalgamation of all protocols in place to protect organisms from contracting, carrying (or) spreading diseases. In nutshell, it is the protective shield of fish (or) shell fish in ornamental realm from infections (viral, bacterial fungal (or) parasitic) agents. The key elements of biosecurity are:

Reliable source of stocks

Adequate detection and diagnostic methods for excludable diseases.

Best management practices

Disinfection and pathogen eradication methods

Practical and Acceptable Legislation

The biosecurity program should be designed based on the needs of the specific site, the fish species to be cultured, life stages to be grown, business needs of the operation and the disease profile of the surrounding region. Overall, a biosecurity program would include, but not be limited to, practices and procedure involving;

Surveillance for the presence of disease organisms,Vaccination, Quarantine and restricted access, Appropriate practices of fish husbandry, Disinfection and Disease treatment (including eradication).

The chain of custody in the ornamental aquatic industry starts with the collection of specimen from the wild. Some of these animals will go up the chain and will eventually end up in the aquarium; others will be used for breeding purpose in farms. In many of the cases, more than two middlemen are involved between the collector/breeder and the exporter some of them stock the animals for some time, some just purchase and resell them completely unpacked. Usually the middlemen are responsible to take care of the animals during transportation.

General security precaution should be carried out in each step of the process to help the activities of both disease prevention and disease control. A standard operating procedures (SOP) manual should be assembled to provide a set of standard rules for biosecurity measures and disease monitoring. This manual should also include things like facility design, facility flow for both personal and stock, rules for restricted or limited access to facility, required visitor log book, disinfection procedures for personal and equipment, a waste management plan, pest control guidelines, and general husbandry and management procedures. Thus biosecurity is a team effort, a shared responsibility, and an on-going process to be followed at all times. From the breeder to the hatchery, to grow out operators, biosecurity measures and good aquaculture practices have to be observed to contribute to the success of the industry.

Legislation

Several procedures and guidelines developed by different agencies, organization or nations deal with the components of biosecurity issues and plans. The common objectives include aspects of protecting animal population, environment, food and the humans itself. Many instrument falling under the terms such as polices, codes agreements, plans, convention, regulations and treaties have been made to achieve the objectives of biosecurity. Examples are given in the table.

International or multinational policy instruments containing elements pertinent to aquaculture biosecurity. Dates are years of initial adoption (from scarfe,2003)

Lead organization	Title
World trade organization (WTO)	Agreement on the application of sanitary and phytosanitary measures (SPS agreement), 1995 convention on biological diversity (CBD), 1992, and its Cartagena protocol on biosafety, 2000

[Table Contd.

Contd. Table]

Lead organization	Title
Food and agriculture organization of the united nations (FAO)	Organization of the United Nations (FAO) codex Alimentarius (codes of hygienic practice for the products of aquaculture), 1981-1999 code of conduct for responsible fisheries, 1995 code conduct for the import and release of international plant protection convention (IPPC), 1997 international council for the exploration of the sea (ICES) code of practice on introduction and transfer of marine organisms, 1994
International maritime organizations (IMO)	Guidelines for control and management of ships' ballast water to minimize the transfer of harmful organisms and pathogens, 1997
United nations (UN)	Biological weapons and toxins convention,1972
International union for the conservation of nature	Guide to designing legal and institutional frameworks on alien invasive species, 1999

Quarantine

Quarantine as per OIE is a process of maintaining a group of aquatic animals in isolation with no direct or indirect contact with other aquatic animals in order to undergo observation for a specified length of time and, if appropriate, testing and treatment, including proper treatment of the effluent waters (OIE)

Quarantine programs for aquatic organism typically involve protocols for inspection of animals for disease agents and certification i.e. the issuing of a stating that a particular lot of animals or a production facility has been inspected and is free from infection by a particular pathogen(s).

Quarantine programs may be effective at several levels. At the international and national levels, these programs form an integral part of strategies aimed to protect the natural environment and native fauna from the deleterious impacts of exotic species. Within the country, quarantining helps to reduce the spread of pathogens, while at the local level, i.e., hatchery and farm level, quarantine of fry and brood stock from an outside source helps to protect the hatchery/ farm potentially devastating losses caused by disease. Quarantine facilities need to be built up at exporter's as well as importer's premises. Facilities also to be built up at all ports of entry.

The stringency of quarantine applied should be commensurate with the estimated level risk. Quarantine procedures, including observation for clinical sings of disease and diagnostic testing, can be conducted in the country of origin, in a country of transit and /or in the receiving country.

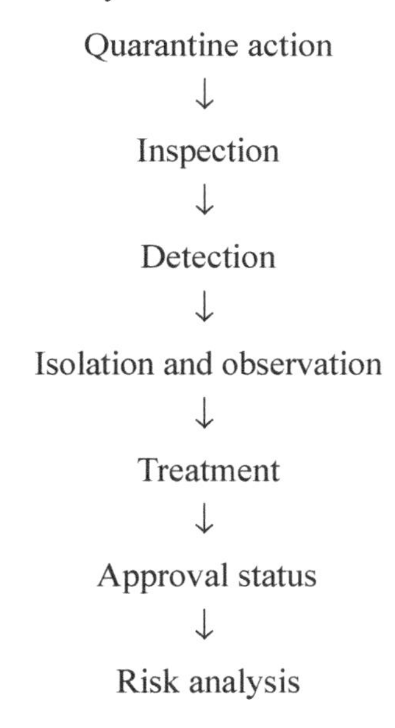

The basic risks to be taken care are

- Pathogens present in natural waters
- Combining animals from different sources
- Water used in the holding facilities
- New species entering the breeding facilities
- Material equipment and plants
- Food supplied to animals
- Personnel
- Replacement of water during transport
- Unwanted animals in holding facilities
- Pathogens possibly present in the natural water

SVC is the only notifiable disease, relevant to our industry, for which some countries have declared themselves disease free, or have implemented on eradication and control program.

Captive bred aquatic animals may carry unwanted pathogens. This is especially dependent on the biosecurity measures of the breeder. The animals may not show any sign of a disease as, in general, captive bred animals are kept in good quality water, fed with proper food, under conditions suited to the species concerned. As a result pathogens may also wander around unnoticed in breeding facilities.

Combining Animals from different Sources

Combining animals from different sources may be by far the most underestimated sources of pathogens in our industry. In fish holding facilities of the collector fish from many different sources are temporarily stored. Fishes collected in separate tributaries may carry different loads of pathogens. the individual risk of each fish being a vector for a pathogen should be added up for multiple localities and per water system. This model seriously increases the total risk of presence of unwanted pathogens. This also applies to fish from different breeders

Water used in the Breeding / Holding Facilities

Another important possible source of infection with unwanted pathogens is the water supplied to the animals. In some parts of the world, water used for aquaculture or storage of ornamental aquatic animals can be quite polluted and with a high risk of presence of pathogens. Such water will undoubtedly infect the animals kept in such system even if the animals are healthy. Again if threshold densities for the diseases are reached than the susceptible animals will get sick, and if not, the susceptible or vector species animals will be able to carry the pathogens up chain of custody.

New Species Entering the Breeding Facilities

Every new species entering a breeding facility poses a risk to the stock already present in the facility

Food Supplied to Animals

In the ornamental aquatic industry, live foods are considered to be healthy and very beneficial to the colors of the animals. Hobbyists often collect live mosquito larvae, daphnia and Cyclops from natural sources. Professional breeders apply artemia and tubifex in big quantities, but also moina and other live foods.

Personnel

The ornamental aquatic industry is labor-intensive industry. In tropical countries as well as in the, western world, person will get in touch with a wide variety product, which may contain pathogens while walking to work people can go into contact with waterborne pathogens on wet roads or from one farm to another. Some pathogens may be present in bird or lizard droppings can be handled unnoticed. Transmittance of disease by and between people are numerous, e.g., personnel movements between farm during coffee breaks, business transaction, etc.

Health Certification

It is a document declaring the health status of aquatic animals in a batch/ consignment, from a country, zone or aquaculture establishment. It is issued by a competent authority of the exporting country for international trade in live or dead aquatic animals products. For internal trade in live or dead aquatic animals , the head of a disease diagnosis service, head of government lab can issue the required certificate. The format and diagnostic tests as specified by the OI's aquatic animal health code and manual of diagnostic tests for aquatic animals may be used for examining the animals and issuing health certificates.

Disease Surveillance, Monitoring and Reporting

In health management, the most current information on the disease is extremely important. Hence, a systematic approach to gather and interpret disease information within the country needs to be developed. For the purpose, we need to establish a networking mechanism to facilitate collection of disease information from all the available sources. Disease surveillance and reporting forms the basis for all national disease control programs. It helps to meet trade requirements and forms the basis for risk analysis. Surveillance includes both passive and active or targeted surveillance. One of the first steps in surveillance is to establish a national disease database to aid regional and international reporting, develop disease control strategies, and identify and prioritize areas for research and capacity building. The national bureau of fish genetic research (NBFGR), Lucknow has initiated steps towards establishing a database for India. The FAO guidelines suggest that countries should provide accurate and timely reporting of disease notification and associated epidemiological information on a quarterly basis to the world organization for animal health (OIE) or to other disease reporting system (e.g. the NACA/FAO, quarterly aquatic animals disease reports QAAD (Asia and Pacific region). To enable this, necessary diagnostic and reporting procedures should be established. Standardized field and laboratory methodologies and resources materials should be developed and field personnel should be trained in disease recognition and reporting to ensure accurate and rapid pathogen identification.

Zoning

Zoning can be highly effective means to restrict the spread of important pathogens of aquatic animals which also aid in their eradication, where a serious disease is present in part of a nations territory and eradication is not feasible, countries should consider the possibility of zoning as a means to eastablish and maintains zones free of the disease and to permit international and domestic trade in live aquatic animals originating from these zones.

Emergency Preparedness and Contingency Planning

A contingency management plan has to be prepared well in advance so that at everyone in the framework known their responsibilities and actions to be taken in the face of an emergency such, emergency planning for aquatic animals' disease should form a core function of government services. In order to respond rapidly and effectively to contain and eradicate serious disease outbreaks caused by transboundary aquatic animal disease (TAADs) and thus minimize their social and economic impacts, countries should develop and test national contingency

Advantages of Biosecurity

There will be meager loss in entire supply chain. This will not only save money but increase the reputation of the trade

It leads to less usage of medication in the steps of the chain and therefore easy to comply with national legislation or use of medicines. Less medication leads to less resistance of pathogens and of course saves money. The measures will lead to better quality and less time needed for recovery from transport the end result will be better position in the competition for customers and will even allow somewhat higher selling price.

Disadvantages

- It demands investments at every level.
- It requires initial setups maintenance and continuous measures
- It will involve time-consuming procedures, the purchase of more expensive equipment, investment in better education of staff and more.

CHAPTER 10 Economics of Livebearers Fish Farm

The art of rearing the ornamental fishes has been disseminated throughout the world. This fascination to rear the vibrant coloured species, has added colours to the tremendous ornamental fish business. As the result, many countries in the Asia and Europe started capturing and culturing the vibrant and fanciful breeds of the fishes. The myriad variety of ornamental fishes has been domesticated and popularized for the business purposes. The establishment of more and more ornamental fish farms and hatcheries made the business people invest in this realm.

The small-scale backyard ornamental fish culture and breeding could largely satisfy only the needs of the hobbyists. In order to undertake the ornamental fish culture in a commercial way and promote export and earn considerable foreign exchange, the advanced technologies should be undertaken to enhance the production level.

However, to make ornamental fish culture a profitable trade, it is necessary to have a fair knowledge on the economics of its farming. This will help the entrepreneurs in their decision-making process. The prime objective in planning and budgeting in a business is maximization of the profit.

Cost

The concept of operational costs will be more appropriate in the economic analysis of enterprises like ornamental fish culture. The operational cost includes fixed cost and variable cost. Variable costs like feed vary with the level of output.

Returns

The output price gives total returns. It is the indicator of profitability of an enterprise.

Risks

Risks are part of the ornamental farming.

Deterministic models will not take the risks into consideration and so the analysis is unrealistic.

The stochastic models include the uncertainty and risks; therefore, it is a reliable model.

S. No.	Item	Description	Backyard Unit (Rs)	Medium scale Unit (Rs)
A	**Capital Cost**			
A1	Cement tanks	Cement tank including storage tanks	120000	200000
A2	Shed cost	Structure with cemented, brick wall, asbestos/ metal/ RCC and Plastic greenhouse with roll up sides, heat and ventilation in hilly areas	–	103000

[Table Contd.

Contd. Table]

S. No.	Item	Description	Backyard Unit (Rs)	Medium scale Unit (Rs)
A3	Live feed facility and feed maker	Cement Tanks/ FRP Tanks, Glass tanks for stock culture	–	65000
A4	Glass tanks	Aquarium tanks including stand	40000	100000
A5	Water supply items	Water line pipes, Motor & Pumps, hose &it's fitting	20000	20000
A6	Electrical items	Wiring material, lightings and its fixtures, submersible heaters etc.	40000	150000
A7	Water treatment equipment	Biological filters, carbon filters, RO units etc.	20000	150000
	Total	**Capital Cost**	**240000**	**788000**
B	**Operational Cost**			
B1	Brood stock fish		12000	35250
B2	Feed		9375	50600
B3	Labour Cost	@ Rs.200 per person x 180 days	36000	36000
B4	Power & Fuel		6250	20650
B5	Packing & Transport		9375	15500
B6	Miscellaneous	Medicine, stock solutions for live feed culture etc.	7000	20000
	Total Operational Cost		**80000**	**178000**
	Total Cost involvement (A+B)		**320000**	**966000**

Production of live bearers in backyard unit

Sword tail – 100 (young ones)

1 female sword tail → 100(young ones) x 4(months) = 400 numbers

For a year [three cycles] = 400 x 3 = 1200 numbers/ year

Mortality at 10 % = 120 Numbers/ year

Remaining Fishes → 1080 numbers / year

100 Female sword tail → 100 x 1080 = 108000 numbers / year

Cost of Selling at rate of 10 per young one → Rs10,80,000

Guppy – 50 (young ones)

1 female Guppy → 50(young ones) x 4(months) = 200 numbers

For a year [three cycles] = 200 x 3 = 600 numbers/ year

Mortality at 10 % = 60 Numbers/ year

Remaining Fishes → 540 Numbers/year

100 Female Guppy → 100 x 540 = 54000 numbers / year

Cost of Selling at rate of 10 per young one Rs5,40,000

Molly – 100 (young ones)

1 female Molly 100(young ones) x 4(months) = 400 numbers / year

For a year [three cycles] = 400 x 3 = 1200 numbers / year

Mortality at 10 % = 120Numbers/year

Remaining Fishes → 1080 Numbers/year

100 Female Molly → 100 x 1080 =108000

Cost of Selling at rate of 5 per young one 5,40,000

Platy – 50 (young ones)

1 female Platy → 50(young ones) x 4(months) =200 numbers

For a year [three cycles] = 200 x 3 = 600 numbers/year

Mortality at 10 % = 60 Number/ year

Remaining Fishes = 540 Numbers/year

100 Female Platy → 100 x 540 = 54000 numbers / year

Cost of Selling at rate of 10 per young one → Rs 5,40,000

Summary

Sum of total production of livebearers 10,80,000 + 5,40,000 + 5,40,000 + 5,40,000 = 27,00,000

Annual profit → 27,00,000 – 3,20,000 = Rs.23,80,000

Monthly profit = Rs.1,98,333.33

References

Ahilan, B., Felix, N., and Santhanam, R.2008. *Textbook of Aquariculture.* Daya Publishing House. New Delhi. P. 150.

Alagappan Mand Vigula K. 2004. Aquarium Fish Breeding Techniques, Fishing Chimes, 24 (5): 26-27.

Anon. 1998. Training Manual on Culture of live food organisms for Aqua hatcheries Central Institute of Fisheries Education, Mumbai.

Anon. 1999, How to develop ornamental Fish farming. Seafood Export Journal Vol 30 (2): 31.

Axelrod, H.R. and Schultz, P.L., 1983. Handbook of Tropical Aquarium Fishes. T.F.H. Publications, Hong Kong, p. 28 30.

Bhat, B.V., 2008. Export oriented aquaculture India: An overview. Fishing Chimes, 27 (10/11): 51-58.

Boyd, C.E., 1992. Water Quality Management for Pond Fish Culture. Elsevier Science Publishers, Netherland, p. 317.

Chairs Andrew. 1986. Guide to Fish Breeding, Interpet Ltd., U. K.

Claude E. Boyd, Criag S. Tucker., Pond aquaculture water quality management. Springer science, p175-200.

Dakin, Nick. 1992. The Book of Marine Aquarium. Salamander Book Ltd., U. K.

Dawes, John.1995. Live bearing fishes- A guide to their aquarium care, biology and classification. Sterling publications, New York, p.120.

Dey, V. K., 1993. Ornamental Fishes. Marine Products Export Development Authority, Kochi. pp. 7-10.

Dey, V. K., 2008. Global Trade in Ornamental Fish: Trends, Prospects and Issues. Abstract, International seminar on Ornamental fish breeding, farming and trade, Cochin, India, pp. 2.

Dholakia A. D. 2005. Economical strengthening with the help of Aquarium fish breeding. Radio Talk at All India Radio, Rajkot, March 2005.

Dholakia A. D. and A. Y Desai. 2004. Aquarium and its maintenance (in Gujarati). Folder published by College of Fisheries, Junagadh Agricultural University, Veraval.

Dholakia.A.D., 2009.Ornamental fish culture and Aquarium management. Daya Publishing house, New Delhi, p.46-88.

FAO, 2007. Fishery Statistics, Aquaculture Production, 2005.

FAO, 2020. Fishery Statistics, Aquaculture Production, 2018.

Felix. S., Anna Mercy. T.V., Saroj Kumar Swain., 2013. Ornamental aquaculture technology and trade in India. Daya Publishing house, New Delhi, p.115-149.

Hossain, S., & Heo, G. J. 2021. Ornamental fish: a potential source of pathogenic and multidrug resistant motile Aeromonas spp. *Letters in Applied Microbiology*, *72*(1), 2-12.

Jain, Atul Kumar. 2005. Aquarium Hardware Course Manual-winter school by ICAR 8-25 Feb: 49-56.

Lochmann, R.T. and Phillips, H., 1994. Dietry protein requirement of golden shiners (*Notemigonus crysoleucas*) and goldfish (*Carassius auratus*) in aquaria. Aquaculture, 128: 277-285.

Moyle, P.B. and Cech, J.J., 1988. Fishes: An Introduction to Ichthyology, 2nd edition, Prentice Hall, Englewood Cliffs, NJ.

MPEDA, 2000. Statistics of Marine Products Exports, 2000. MPEDA, India, p. 25.

Peter, W. Scott. 1987. A Fish keeper's guide to live bearing fishes. Salamander Book Ltd., London, U. K.

Petrovicky, I., 1993. Tropical Aquarium Fishes. Chancellor Press, London, p. 258.

Purdom, C.M., 1993. Genetics and Fish Breeding. Chapman and Hall, London, p. 277.

Radha, C. Das and Archana Sinha, 2003. Ornamental fish Trade in India (West Bengal and Tripura). Fishing Chimes, 23 (2): 16-18.

Santhanam, R. N. Sukumaran, and P. Natrajan. 1999. A manual of Freshwater aquaculture. Oxford and IBH Publishing Co. Pvt. Ltd, New Delhi.

Saroj, K. Swain and Partha Bondopadhyay, 2002. Breeding Technology in Ornamental Fish. Fishing Chimes 22(3): 56-60.

Sexena, Amita. 2003. Aquarium Management. Daya Publishing House, New Delhi.

Sharma, L. L. 2005. Aquarium Plants and their culture. Course Manual-Winter School by ICAR 8-28 Feb: 14-16.

Sharma, L. L., S. K. Sharma, B, K. Sharma and V. P. Saini 2005. Tropical Freshwater aquarium fishes. Course Manual-Winter School by ICAR 8-28 Feb: 4-7.

Sharma, Omprakash. 2005. Food and feeding of Aquarium Fishes. Course Manual Winter School by ICAR, 8-28 Feb: 86-88.

Srivastava, C. B. L. 2002. Aquarium Fish Keeping. Kitab Mahal Allahabad.

Swain, S. K., J. K. Jena and S. Ayyappan. 2000. Prospects of Freshwater Ornamental Fish culture in India with reference to export marketing. Fishing Chimes 20 (10 and 11): 99-101.

Varma, S. K. 2005. Freshwater ecosystem and practices in the ornamental fish aquarium management. Course Manual-Winter School by ICAR, 8-28 Feb: 57 60.

Subject Index

B

C

G

H

I

J

L

M

N

O